METHANOL

Bridge to a Renewable Energy Future

John H. Perry, Jr.
Christiana P. Perry

UNIVERSITY
PRESS OF
AMERICA

Lanham • New York • London

Copyright © 1990 by
University Press of America®, Inc.
4720 Boston Way
Lanham, Maryland 20706

3 Henrietta Street
London WC2E 8LU England

Library of Congress Cataloging-in-Publication Data

Perry, John H.
Methanol : bridge to a renewable energy future /
John H. Perry, Jr., Christiana P. Perry.
p.   cm.
Includes bibliographical references.
1. Alcohol as fuel. 2. Methanol.
I. Perry, Christiana P.   II. Title.
TP358.P47   1990   333.79'3—dc20   89–24978 CIP

ISBN 0–8191–7659–1 (alk. paper)

The paper used in this publication meets the minimum requirements of
American National Standard for Information Sciences—Permanence
of Paper for Printed Library Materials, ANSI Z39.48–1984.

Dedicated

To

A Better Understanding Of Our Planet

And

Its Resources

ACKNOWLEDGEMENTS

We would like to thank all who assisted in carrying this project to its completion, especially recognizing the contributions of the following:

Jack Silvey, former Deputy Assistant Secretary for Policy at the U.S. Department of Energy, who edited the manuscript and helped prepare it for publication.

Abe Persons and his associates at Perry Energy Systems, whose contribution to the development of the SSP process inspired the production of the book. Mr. Persons prepared that portion of Chapter Six devoted to Seafuel^R, and provided valuable advice throughout the writing of the book.

We would also like to express appreciation to:

T. Nejat Veziroglu, Professor of Mechanical Engineering at the University of Miami and director of the Clean Energy Research Institute, for his support and the use of his color photographs.

Peter Hoffmann, editor of the *Hydrogen Letter*, for his interest and update on methanol-hydrogen technologies.

All the staff at Perry Oceanographics, Inc. who helped in the preparation of the manuscript, with special thanks to Betty Winter.

TABLE OF CONTENTS

FOREWORD

This book should be read by any citizen interested in energy and environment problems. John and Christiana Perry have made an historic contribution to the literature and understanding of what is happening in our world in this area. I have long been fascinated with the experiments, innovation and creative thinking of this father/daughter team. It has been a privilege for me to rely on their expertise regarding many issues that I have dealt with in the United States Senate.

As a Member of the United States Senate Subcommittee on Commerce, Science, and Transportation, it is incumbent on me to keep abreast of developments in the field of energy and how it relates to the security of the United States and world environment. I therefore took an immediate interest in reviewing what John Perry and his daughter Christiana have to offer as solutions to these two vital problems. I have witnessed the work being done in their plant in Florida, the patents given by the U.S. Patent Office and the background of Mr. Perry in the field of oceanography over the past twenty-five years and I must say that this book should be not only read, but studied by those in appropriate positions in the Congress and the administration so that they may be able to make the right decisions regarding these important matters.

Our future as a nation and our role in world leadership is of vital importance to every citizen, and this book does an excellent job of documenting the facts and solutions to what is becoming an increasingly critical issue.

U.S. Senator Larry Pressler
South Dakota

PREFACE

This book has been written to raise the public's awareness of two conditions that threaten both the stability of the world's economies and our current quality of life. The first of these is our extreme dependence on oil to fuel our economic engine--even knowing the risks of disruption to that vulnerable source of energy. The lessons of the 1970s have been forgotten in the past few years as oil imports have increased dramatically, and this growing dependence on cheap oil has once again placed the majority of the world's economies in positions of great risk from the threat of a cut-off in oil supplies.

The second condition is the severe damage to our environment that is occurring every day simply due to our dependence on all fossil fuels (oil, gas, coal). This damage is immediately visible when you drive through one of our major cities in the summer, penetrating the yellowish-brown cloud that engulfs it. Less visible but more damaging are the impacts of acid rain on our forests, lakes and streams. Perhaps most damaging are the potential impacts of trends toward global warming, also largely a product of fossil fuel combustion.

The authors believe that both of these conditions can be addressed, and mitigated, through efforts to develop alternative fuels; and the fuel that we believe to have the greatest potential for both immediate and longer-term gains is methanol.

It is important to realize when reading this book that any energy system has its benefits and drawbacks. Consequently, any one plan that promises unconditional rewards should be approached with a high degree of skepticism. Solar power, the ecologist's dream, is seen as flawless from an environmental standpoint; yet in its present state of development, poor economics have limited its application to certain specialized markets. Similarly, early projections of low-cost nuclear power as the long-term solution to our electricity needs have been radically altered as major safety and reliability problems have developed and costs have sky-rocketed.

In our discussion of the potential for methanol as an effective substitute, we are not attempting to offer a panacea. We have tried to address as objectively as possible the various degrees and types of benefits that can be derived from increased use of methanol, first as a supplementary transportation fuel and later as a bridge to a more advanced hydrogen economy. We ask the reader's indulgence in instances where our belief in the benefits of this fuel may appear to have colored that objectivity.

John H. Perry, Jr.
Christiana P. Perry

INTRODUCTION

In 1987, the United States economy consumed 76.8 quadrillion Btus (Quads) of energy. Of that amount, 68.5 Quads, or 89%, were produced from fossil fuels (oil, natural gas, coal). Oil and oil products comprised almost half of fossil fuel use, and the trend was definitely toward increased consumption of all three fuels.[1]

These statistics are quite sobering when consideration is given to the economic and environmental lessons that the U.S. and the world have learned over the past two decades. We have experienced oil price shocks that first tripled oil prices and doubled them again within five years, fueling the highest level of inflation experienced in this century and causing substantial reductions in the growth of our economy. The nation's memory is short, and the recent return to low oil prices has sparked a rapid growth in demand and increasing dependence on imported oil. We are rapidly approaching oil import levels that exceed our domestic levels of production; and even though current prices are low, our 1987 imports cost the economy about $40 billion, fully one-quarter of the annual trade deficit for the year.

This period also has been one of increasing focus on world environmental problems. The two issues with the highest level of interest and activity in the U.S. have been air quality and acid rain. Both are caused by the combustion of fossil fuels. Several U.S. cities are facing restrictions on their increased growth as a result of failure to comply with the Clean Air Act limits on emissions of air pollutants, both from mobile and stationary sources. Restrictions on automobile emissions and fuels are being tightened in many urban areas, and older coal-fired utility plants that were not covered in earlier Federal air quality regulations are likely to face new requirements from this session of Congress.

In addition to air quality, a new environmental issue has emerged over the past few years that has far broader implications. Global warming has been identified as a phenomenon that already is underway and may have severe impacts on both the environment and the world's economies as it develops into the next century. The principal cause, again, is the combustion of fossil fuels, creating an expanding blanket of carbon dioxide and other gases that is covering the earth and retaining heat. The science and theories behind such prognoses for the future

---

[1]U.S. Department of Energy/Energy Information Administration, *Monthly Energy Review, October 1988,* (Washington: DOE/EIA, January, 1988), p. 9.

need much further study, as should any projections that attempt to forecast events 50 years into the future with any certainty. However, there are enough indicators of existing and potential environmental problems with fossil energy to reinforce the need for examining options to the growing U.S. dependence on these fuels.

The thesis of this book is that the need for alternatives to fossil fuels in the U.S. will only continue to grow, and the option that has the most immediate near-term applicability to our needs is methanol. The focus is on the transportation sector, both because the need is perceived to be greater there than in the other sectors and methanol's characteristics match that sector's near-term requirements.

This thesis is developed by looking first at the economic and environmental implications of our current fossil-dependent energy economy. The current range of alternative fuels then is examined, followed by an in-depth look at methanol as the prime alternative for the transportation sector. The chapters are developed as follows:

Chapter 1 addresses our growing dependence on petroleum and natural gas, looking at the advantages and disadvantages of each fuel and the potential for severe economic problems in the event of another supply disruption.

Chapter 2 completes the examination of fossil fuels with a look at current and future uses of coal in the U.S. energy mix. This chapter also addresses environmental concerns that are inherent in using all three fossil fuels.

Chapter 3 examines current alternatives to fossil fuel use and identifies which sectors of the energy economy these options best support. Nuclear power and a wide range of renewable energy sources are examined, and one conclusion is that there are few options or incentives to find other fuels for the transportation sector.

Chapter 4 examines the applicability of methanol as an alternative fuel, primarily focusing on the transportation sector.

Chapter 5 addresses the problems of market penetration in the face of cheap petroleum-based fuels.

Chapter 6 addresses the issue of whether methanol could be produced in commercial quantities if a large transportation market did exist.

Chapter 7 takes a longer view, addressing the potential for a hydrogen economy and the bridge to that future that methanol could provide.

CHAPTER ONE

OUR DEPENDENCE ON FOSSIL FUELS

Energy has a special significance in the economy. It is the basis of industrial society. Oil, in particular, carries special importance and thus special risks because of its central role, its strategic character and its geographic distribution--and the recurrent pattern of crisis in its supply. Current circumstances allow us to view the energy scene dispassionately, and to recognize how much has changed. It will be remarkable if we get to the end of the 1990s without our energy security being tested by political or technical crises. Today, however, we do have considerable security margin, and that allows us the hard-earned luxury of addressing the question of energy security in the 1990s and how it can be maintained and enhanced-- not with panic, but with prudence.

-Daniel Yergin

This chapter will examine U.S. use of oil and natural gas, looking at the advantages and disadvantages of each fuel form and our level of dependence on each, and the implications for the U.S. economy of their continued and expanded use. A major concern is the energy security implications of increasing U.S. dependence on both fuels and the potential for economic shocks such as those experienced in the 1970s. Since oil and natural gas compete in the utility and industrial sectors, an increase in oil prices will be reflected rapidly in natural gas markets; and at today's low prices for both fuels, few incentives exist to bring alternative fuels into the market, particularly in the transportation sector. This chapter will explore the key factors that have led to this situation and how these markets are expected to evolve.

OIL

The U.S. Petroleum Economy

As recently as 1950 oil supplied less than one-third of the world's commercial energy, with coal accounting for over half of worldwide energy use (Figure 1-1).[1] With the evolution of refining techniques and the growth of the automobile industry, petroleum became the dominant fuel. It contains more energy per unit of volume than any other major fuel, and it quickly proved preferable to coal for many uses because of its versatility. Easily extracted and transported, it is a versatile fuel that requires relatively modest investments in transportation and technologies for combustion. Accordingly, oil extraction expanded over 400% from 1950 to 1973, its increased use in turn fueling most of the increase in energy consumption during the same period.[2]

Free world energy use (excluding biomass and other traditional fuels) has grown at an annual compound rate of over 3%, more than tripling since 1950.[3] Petroleum has accounted for an increasing share of this ever growing demand. Today, the world gets nearly half of its energy from oil, with the U.S. owing more than 40% of its energy use to this fuel (Figure 1-2).

Figure 1-1

Global Energy Use, 1950-1985*

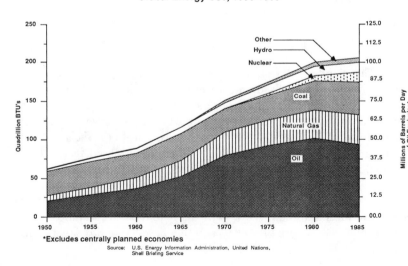

*Excludes centrally planned economies

Source: U.S. Energy Information Administration, United Nations,
          Shell Briefing Service

Figure 1-2
Energy Requirements by Fuel Type
1985

World

211 Quadrillion BTU's

US

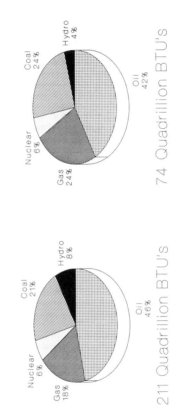

74 Quadrillion BTU's

Source: Energy Information Administration

Within the U.S., oil use is concentrated primarily in one sector: transportation (Figure 1-3). While total U.S. consumption of oil for 1985 was 31 quads of energy, or approximately 15.5 million barrels per day (MMB/D), 9.2 MMB/D were consumed in the transportation sector.[4] This amounts to 63.5% of total U.S. oil use, a quantity greater than the U.S. produces, placing the U.S. in a historically unique position (see Figure 1-4). In 1987, use of oil for transportation alone was about 9.8 MMB/D, and that represented an amount equal to 117% of total U.S. petroleum production.[5] In addition, U.S. vehicles are more than 98% dependent on petroleum based fuels,[6] and this heavy concentration of oil use in the transportation sector suggests that a remedy for our heavy dependence on oil could lie in the exploration of alternate transportation fuels.

As the demand for oil has grown, the abundance of low-cost reserves has decreased. The U.S., a pioneer in oil use, reached its peak production capacity in 1970 with 11.3 million barrels per day.[7] Since that time, spare capacity has vanished and production has declined. Declining domestic production, coupled with an increase in consumption, has resulted in an increase in imports. The U.S. has been a net importer of energy since the 1950s, when coal was displaced by petroleum and our volume of coal exports dropped. Although the U.S. holds only 5% of the world's proven economically recoverable oil reserves, it is still the second largest single producer of oil in the world, as well as the largest consumer.[8]

With the sharp drop in oil prices in 1986, U.S. oil imports have increased considerably; and recent projections agree that sometime in the 1990s, the U.S. will import more than half of the oil it consumes. The worst-case scenario, envisioned by the National Petroleum Council in a March 1987 report, predicted that during the next decade, U.S. dependence on foreign sources will range from 48 to 60% of its oil needs.[9] While all projections are not this high, there is general agreement that U.S. imports will increase significantly in the next ten years. The Department of Energy's 1987 *Energy Security* report to the President has projected that imports will rise to between 8 and 10 million barrels per day (MMB/D) in the 1990s, a level that is half or more of the nation's oil consumption.[10]

Recent trends appear to support the hypothesis that this may occur sooner rather than later. Total petroleum imports, which in 1985 were 5.1 MMB/D, jumped sharply to 6.2 MMB/D in 1986, increased again to 6.5 MMB/D in 1987, and are averaging 7.1 MMB/D for the first ten months of 1988. The monthly levels for the last quarter of 1988 were very close to

Figure 1-3

U.S. Primary Energy Consumption   Oil Consumption By Sector

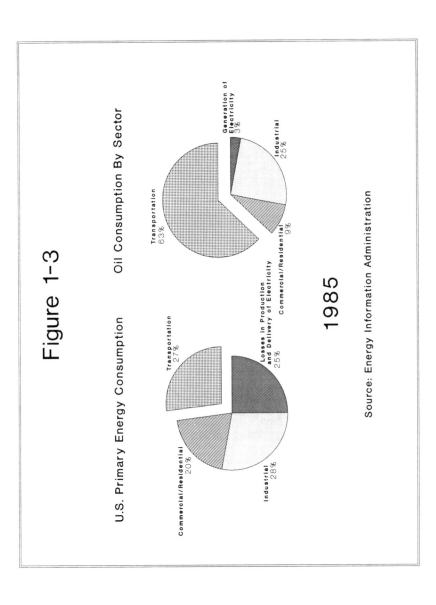

1985

Source: Energy Information Administration

Figure 1-4

Transportation Consumes More Oil Than U.S. Produces

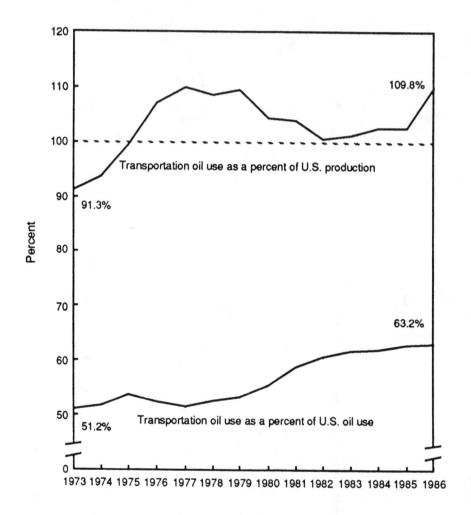

Source: Energy Information Administration

the 8 MMB/D projection.[11] At this rate, the U.S. could become a net importer of oil much sooner than expected, a milestone that no one is anxious to achieve.

While no one can predict future oil prices with any degree of certainty (the past decade of oil price projections proves this point), key elements that drive the long-term supply and demand picture for oil are not improving. The amount of worldwide proven, economically recoverable reserves is not increasing; but the demand for oil is. Furthermore, with the drop in oil prices, the movement toward energy conservation (and reduction of oil demand) has slowed dramatically. Finally, other variables, such as the rapid rise of energy demand in newly industrializing countries, where needed economic growth is dependent on low capital and operating costs, contribute to an increase in the demand for oil. In short, the only certainty about oil is that without a major price shock, demand will not drop anytime soon.

Oil and U.S. Energy Security

In 1986 oil prices plummeted to half of what they had been only two years earlier, reaching their lowest levels in a decade. While prices have risen somewhat since then, the dissension within OPEC over production quotas for oil has given some respite from high oil prices that may extend for a while longer. Amidst a background of low oil prices, it may seem superfluous to advocate the active development of an alternate fuel. It is precisely such a time, however, when we can best afford to act. The experience of the 1970s showed that decisions made in a crisis environment are not well-designed to provide long-term solutions to complex problems, such as those faced by the U.S. in its energy markets. Effective long-range planning is better accomplished in a period of calm; and it is now that the U.S. should take preventive measures against future energy shocks and make an investment in energy self-sufficiency.

In order to make rational judgements on planning for an energy future that is not so dependent on oil, it is necessary to review the events of the past two decades that have brought us to today's energy markets. This includes the two "energy crisis" experiences in the 1970s, which took the U.S. and the rest of the world on an economic roller coaster ride in energy prices. (Figure 1-5 shows the historical relationship between international events and oil prices.)

The term "energy crisis" originated in the seventies when the 1973 Arab oil embargo caused prices to skyrocket and the world was forced to look for alternative energy sources. A tight oil market was hit with a relatively small disruption in available supplies, and the price of oil more than tripled over night. A similar shock occurred in 1978-79 with the Iranian Revolution, where loss of a small fraction of world oil supplies caused a tight market to more than double the world price of crude oil.

Figure 1-5

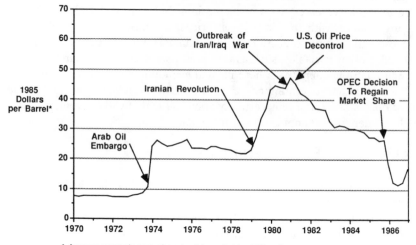

Oil Prices Reflect International Events

* Average quarterly cost of crude oil imported by U.S. refiners.

Source: U.S. Energy Information Administration

It is important, however, to further qualify the definition of the term "energy crisis." First, the U.S. has not faced an actual "energy" crisis, that is, the inability to get adequate energy supplies from the market. Even oil was available in desired quantities during the embargo, for those willing to pay the price. What the U.S. did experience was the economic shock of having a particular source of energy increase sharply in price. Furthermore, many forms of energy other than oil are available--forms that, at a cost, are not susceptible to such disruptions. The simple fact is that the U.S. has developed a strong dependence on oil because of its low price for most of this century and its flexibility and ease of use in a variety of forms, most importantly as a fuel in the transportation sector. This dependence developed over a long period of time, and it will take time and effort to reduce our need for oil.

In the decades of the 1950s and 1960s, U.S. industries and the public found petroleum products to be flexible, available, and low-cost sources of energy. Imports from the Middle East were very cheap, less than $2 per barrel in the late 1960s, and many industries, commercial users and individual home-owners converted to oil during that period.

This was particularly evident along the eastern seaboard of the U.S., where oil became the predominant fuel for everything from electricity generation to home heating. For the wide expanses of the middle and western U.S., the automobile provided low-cost, flexible transportation that almost everyone could afford. It was not unusual in the '50s and early '60s to find service stations conducting "price wars" to expand business, where gasoline sold for as little as 15 cents per gallon.

Once an initial investment has been made in energy capital stock, whether industrial or residential, it is a long-term proposition to replace it, particularly when it may yet be early in its useful life. A home furnace may last 15 to 20 years or more, and an industrial boiler can be serviceable for more than 30 years. The Long Island Lighting Company, which has been trying for several years to bring the completed Shoreham nuclear power plant into service, remains 100% dependent on oil for the generation of electricity, 15 years after the Arab oil embargo.

With oil still playing such a major role in our energy mix, the implications of our growing dependence on foreign imports of this commodity are clear. The most serious of these is our increasing vulnerability to the energy and economic effects of another oil supply disruption.

Almost two-thirds of the world's recoverable reserves of oil are located in the Persian Gulf, and nearly a quarter of free world oil supplies is produced and shipped from this region. Older producing fields will continue to decline, such as those in the U.S.; and the share of the world market that is supplied by Persian Gulf oil will continue to grow, simply because of the overwhelming volume of low-cost reserves that are located in the region.

Recent events, such as the Iran-Iraq war, underscore the tenuous nature of the supply lines from the Persian Gulf; however, the world's dependence on energy supplies from the region means that the political stability of the Gulf states cannot be ignored by any of the major world powers. The U.S. has come to regard a steady flow of oil from the region as a cornerstone of its foreign policy. This policy is acknowledged in the Department of Energy's *1987 Report to the President:*

> *All recent Administrations have viewed the strategic Persian Gulf region as an area of vital security interest. This is evidenced by the creation of the U.S. Central Command (a unified command assigned to deter or oppose Soviet aggression in Southwest Asia and the Middle East) and by regular presence of U.S. Navy vessels in that part of the world.*[12]

The reason for such a costly "insurance policy" is simple. To restate the point again, the Middle East is, and will be, the world's major supplier of crude oil. Every day over one-third of the world's oil trade, roughly between 6.5 and 8.5 million barrels, passes through the Straits of Hormuz.[13] While most of this currently is destined for Europe and Japan, the U.S. receives about 20 percent. Moreover, as recently as the early 1980s, when the international volume of oil trade was higher, almost half of the world's oil passed through the Gulf.

It is important to note that the magnitude of U.S. imports is not a major factor in the price calculation. The market for oil is truly a world market; and barring the return to a price-controlled oil market in the U.S., the world prices for oil will be the prices charged by U.S. producers, regardless of costs of production. The only effective means of reducing the price impacts of oil market fluctuations is to reduce demand. The overall tightness (available supplies compared to demand) of the world oil market is, however, the major factor in what will happen to oil prices; and evidence of trends toward a tighter oil market are the basis for current concerns with prices in the decade of the 1990s.

A brief picture of the magnitude of the world's reliance on Persian Gulf states for oil supplies may bring these issues into sharper focus. Nearly two-thirds of the free-world's oil reserves lie in the five countries surrounding the Gulf (Figure 1-6). Saudi Arabia leads the list, with 27% of free-world, economically recoverable reserves. Kuwait has 15%, the second largest oil reserves held by a single country, while Iran and Iraq each have 8%.[14] Furthermore, the "reserves to production (R/P) ratio," a key measure of how long oil supplies can be expected to be produced from a region, is skewed heavily in favor of the Persian Gulf countries. Table 1-1 shows that the U.S. is at the bottom of the list, with only 8 barrels of proved reserves for every barrel of oil being produced on an annual basis. The R/P ratio for Saudi Arabia, on the other hand, approaches 100.

Surplus production capacity, the capacity to expand output quickly if demand grows, is also unequally distributed. When OPEC was forced to cut production because of sales lost to others, most of the non-OPEC producers held their production at maximum sustainable levels. As a result, OPEC is best prepared to expand its output in response to a sudden increase in demand; and over two-thirds of the world's 10 MMB/D surplus production capacity lies in the Persian Gulf. The remaining third lies primarily with other OPEC countries--Libya, Nigeria and Venezuela.[15] Because of this combination of production capacity and reserves, Department of Energy has predicted that the Persian Gulf's

Figure 1-6

Estimated International Crude Oil and Natural Gas Proved Reserves, End of Year 1976 and 1986

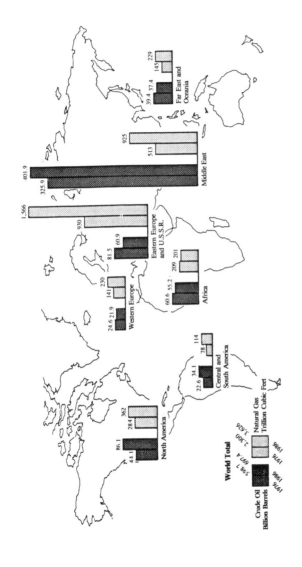

Note: Bars are scaled in proportion to the Btu contents of the reserves. One billion barrels of crude oil equals approximately 5.3 trillion cubic feet of wet natural gas.

Source: DOE/EIA Annual Energy Review 1986

Table 1-1

World Oil Reserves and Reserves to Production Ratio, 1986

Country	Reserves (billions of barrels)	Ratio of reserves to production (n:1)
Saudi Arabia*	169.2	95
Kuwait*	74.6	182
Soviet Union	63.0	15
Mexico	54.7	62
Iraq*	47.1	76
Iran*	48.8	59
Venezuela*	25.0	38
United States	24.6	8
Libya*	21.3	56
China	19.1	20
Nigeria*	16.0	30
United Kingdom	9.0	10
Algeria*	8.8	26
Indonesia*	8.3	16
Canada*	6.9	13
India	3.5	20
Qatar*	3.2	28

*OPEC Members
Source: EIA Data

share of the world oil market, which is under 25% currently, will rise to between 30 and 45% by 1995, and that the market share for all OPEC countries will rise from 40% to between 45% and 60% during this time period.[16] This double dependence--dependence on oil as a principal fuel for our economic well-being and dependence on imports as a major source of supply--places the U.S. economy in an exceptionally vulnerable position.

Economics of Oil Dependence

The United States has a history of self-reliance that has extended to its needs for critical commodities to keep its economy operating. As our dependence on petroleum increased in the '50s and '60s, this situation began to change; but it was not until the 1973 oil embargo that we had any sense of the magnitude of this change in our national economic picture. The physical quantities were enormous. We were using 17.3 million barrels of oil every day, more than 3 gallons per day for every person in the U.S., and more than one-third of that amount was imported.

This revelation of U.S. dependence on one key commodity is, however, one that seems to maintain its significance in the minds of the people only when markets are tight and prices are high. When markets are soft and prices low, as they have been since 1986, public memory is short. It is important to realize that there is a cyclical nature to such a level of dependence, and that our actions during periods when prices are low have a substantial influence on the economic severity of the period when prices again will increase. A brief look at the economics of oil since 1973 will identify some of the key factors that affected prices.

Tight Markets

In the period prior to the 1973 embargo, world oil prices were low, less than $4 per barrel, and U.S. demand was growing rapidly. Domestic production had peaked in 1970; and although the U.S. was producing almost 11 MMB/D, it was importing more than 6 MMB/D. The world market was tight, with little excess productive capacity outside the OPEC nations.

As a result of perceived U.S. support for Israel in the Arab-Israeli War, an oil embargo of the U.S. was announced by the Arab-OPEC nations; and the world supply of oil was cut back briefly by 3 to 4 MMB/D in October of 1973, with an average reduction in world supplies of about 1.5 MMB/D. This cutback was only about 10% of the 31 MMB/D OPEC produced; yet in such a tight market environment, it resulted in the tripling of oil prices from about $4 to $12 per barrel ($7.50 to $25.00 per barrel in 1985 dollars).[17]

Following the initial shock of the embargo, nominal oil prices remained relatively flat for the next few years, which meant that they were dropping steadily when corrected for inflation. U.S. demand for oil dropped by about one million barrels per day for the next two years.

Then, with gasoline lines forgotten, it recovered sharply, exceeding pre-embargo levels in 1976 and peaking in 1978 at a level of 18.8 MMB/D. Imports remained relatively flat immediately following the embargo and then began to grow sharply, peaking at 8.8 MMB/D in 1977.[18] It was only the start of production from the huge Prudhoe Bay field in Alaska that kept U.S. production from falling and imports from going higher.

From November of 1978 to April of 1979, during the revolution in Iran, the second crisis began when 6 MMB/D of Iranian production suddenly was removed from the world market. During this period, Saudi Arabia increased its production to more than 10 MMB/D, offsetting more than half of the loss in Iranian production and leaving a net worldwide supply disruption for the world of about 2 to 2.5 MMB/D for a six-month period. However, the world oil market was very tight during that period; and by mid-1980, prices had more than doubled as a result of this cutback, rising from an average landed cost of about $14 in 1978 to almost $34 per barrel in 1980 ($22 to $44 per barrel in 1985 dollars).[19]

Following the second oil shock, prices remained high; and both U.S. demand for oil and deliveries of imported oil dropped steadily. In 1980 with the onset of the Iraq-Iran war, about 6 MMB/D again was removed from the world market; but market conditions had changed dramatically. High prices had caused sharp reductions in world demand, creating a soft market; and this major reduction in world supplies of oil was barely reflected in oil prices. Many alternate fuels were believed to be on the verge of economic production, and the Federal government invested heavily in this area.

Oil prices peaked in 1981 and began a period of several years where nominal prices were flat or slightly declining, a condition in that period of high inflation which meant real prices were dropping much more rapidly. Programs for the development of alternative fuels had been based on projections of $40 to $50 per barrel of oil, and these initiatives slumped quickly as oil prices fell below $30 per barrel. The 1981 deregulation of oil and oil product prices accelerated the fall in prices. Figure 1-5, presented earlier in this chapter, gives a clear picture of the steady (sometimes precipitous) decline in real prices for oil since 1981. The sharp reductions in Federal support for alternative fuels programs throughout this period, based on the assumption that the free market would decide best which programs should go forward, contributed to the demise of these programs.

Dissension among OPEC partners in 1986 caused additional supplies of oil to reach the market, and the price fell by about 50% in a very short period, stabilizing at about $14 per barrel. U.S. demand increased sharply, as did imports. This rise in imports is leading many to find similarities between current trends and those preceding the shocks of '73 or '79. Oil imports rose sharply, hitting 45% of the total U.S. supply in July 1987, the highest one-month level in seven years.[20]

One measure of the potential for another oil crisis uses the level of excess production capacity as an indicator. OPEC initially gained control of the world oil market when it was using about eighty percent of its production capacity.[21] In 1979, when the second oil crisis occurred after the outbreak of the Iranian revolution, OPEC was using 90% of its production capacity, or 32 MMB/D.[22] This 80% to 90% benchmark is used by many analysts as an indicator of a tight oil market and a greater level of OPEC control. Although OPEC production had fallen well-below the 60% level prior to the 1986 price drop, by August 1987, OPEC was producing at roughly 70% of its capacity. According to a spokesman from the American Petroleum Institute, the U.S. is absorbing nearly 2 MMB/D, or 20% of the excess capacity that OPEC had in January 1986. This amount "gets us one-third to one-half the distance toward the 80% point, . . . and of course, demand is rising elsewhere in the world."[23]

Slack Markets

The threat to U.S. national security and the possibility of another oil embargo are undoubtedly the most dramatic aspects of our dependence on foreign oil. A more subtle aspect of this dependence may be more serious for the U.S. in the long term: the economic costs of our dependence on oil imports.

When oil prices are low, the near-term effects on the U.S. economy can be beneficial in terms of prices. However, low prices also spark increased demand for oil while, concurrently, slowing or stopping critical structural changes to the economy that no longer are economic in the new price environment. This can create effects that are far more severe in the long run. Imports have jumped by more than 2 MMB/D since early 1986, as a result of the drop in oil prices. Low prices have further slowed domestic oil exploration and also have discouraged drives toward alternate energy systems, which already had slowed significantly. Even if there were to be a sharp rise in oil prices, both would suffer from substantial lags in start up time as a result of an extended period of inactivity. The Energy Department estimates a lag time of 5 to 10 years is likely to elapse between each new investment decision and any actual production resulting from it.[24]

The lower price of oil implies more than just the increase in imports. Our domestic production has dropped as the profit margin for oil exploration has decreased sharply. In the Middle East, where excess capacity is easily available, finding costs for oil are virtually irrelevant. But in the U.S., where new reservoirs are usually small and hard to find, low oil prices discourage exploration. While as recently as the 1910-1920s we could find 250 barrels of oil per foot drilled, today we are lucky to find 8 to 10 barrels of oil per foot.[25] Furthermore, while the average U.S. well produces 14 barrels of oil per day, an average well in Saudi Arabia produces 7,000 barrels/day.[26]

The diminishing returns from prospects in the Continental U.S. are forcing oil companies to search farther offshore, in deeper formations, and in harsher climates. As a result, production from known reserves is not being replaced by new finds; and the increase in U.S. oil imports has been accompanied by a decrease in domestic production of about 700,000 barrels per day in the two-year period, 1986-1987 production may continue to fall at a rapid rate if oil prices remain low (See Figure 1-7).

The drop in oil prices has caused a drop in worldwide oil exploration. Exploration and development expenditures by large oil companies in 1986 were down about 30% from 1985. However, this decline was largest in the U.S., (almost 40%) where drilling has fallen to pre-World War II levels.[27] While at the end of 1981, 6,200 oil drilling rigs were in active service throughout the world, this number was down to 3,500 by the end of 1985. As a result of declining prices in 1986, the number fell even further, with only 2,200 rigs still active at the end of 1986.[28]

Within the U.S. the drop was much sharper, from about 4,000 rigs operating in 1981, to less than 2,000 in 1985, and halved again to less than 1,000 rigs operating at the end of 1986. Nationwide, crude oil output has fallen by more than one MMB/D since early 1986, dropping below 8 MMB/D in September 1988, the lowest levels since 1977.[29] Furthermore, the precipitous nature of the 1986 price decline caused the withdrawal from the market of independent operators who cannot withstand the shocks of an uncertain market. This instability, price volatility and the 50% collapse in prices may significantly deter investment decisions that ordinarily would follow a price rebound, causing structural damage in the U.S. oil market.

The reason for the drop in drilling is simple: loss of profitability. The Department of Energy estimates the average cost U.S. companies paid to find oil and gas reserves for the period between 1982-85 to be approximately $8.50/barrel. The marginal direct cost to extract this was another $5.00/barrel, excluding taxes.[30] Furthermore, the 1986 tax reform law will add an estimated additional $10 billion burden to the

Figure 1-7

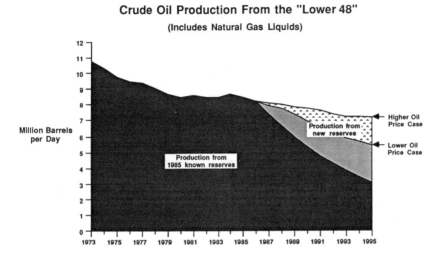

Crude Oil Production From the "Lower 48"
(Includes Natural Gas Liquids)

Source: U.S. Department of Energy - *Energy Security Report, 1987*

domestic industry.[31] Given U.S. exploration and development costs in relation to recent oil prices, it is not difficult to see why exploration and development have slowed dramatically and U.S. oil production has dropped.

Other Economic Effects

Another set of economic costs from oil imports faces the U.S. economy whether the price is high or low. When oil prices are high, some costs to the economy are recognized immediately at the gas pump. But the fact that much of this oil is imported has more serious trade consequences for the U.S. economy. Oil imports are a major contributor to both the balance of payments deficit and the decline in value of the dollar in relation to foreign currencies. In the period from 1976 to 1986, for example, U.S. consumers paid about $640 billion for imported oil, the equivalent of 24% of total U.S. merchandise costs, and an amount virtually equivalent to the nation's trade deficit over that period.[32]

When oil prices are low, the effects on the economy can be equally negative in the short run and quite severe in the long run. The most immediate effect of a drop in oil prices on U.S. trade is an increase in imports. In 1986, although the price of imported oil decreased by $13 per barrel, the U.S. still spent about $75 to $80 million per day on oil imports, and recorded a $29 billion energy trade deficit, accounting for 20% of our total merchandise deficit of $152 billion.[33] The reason this number is so high is that net oil imports increased by 29% or 1.2 MMB/D, bringing our net daily imports to 5.4 MMB/D.[34] This trend has accelerated, as oil imports for 1987 cost the U.S. economy about $40 billion and accounted for one-quarter of the entire U.S. trade deficit during that period.[35]

When oil prices increase faster than inflation, the U.S. has to work harder for each barrel of oil; and payments to foreign governments increase. Higher real prices at home cause a decrease in the demand for goods and services which can lead eventually to an excise tax on the economies of importing countries. An economy facing inflation and balance of payments deficits will suffer from slowed growth and an increase in unemployment. This experience is easily identified in the price hikes of 1973-74, where some analysts estimate the cost to industrial nations from these increases will eventually be over $500 billion in lost economic growth.[36]

Fossil fuels, particularly oil, currently are critical to the smooth functioning of our nation's industrial and transportation infrastructures and to the economy as a whole. But given the sacrifices that are becoming apparent in our heavy dependence on these fuels, it is critical that we not allow this addiction to continue without seriously examining the alternatives. It is neither possible to discontinue using oil altogether, nor is it necessary. It is, however, vital that we explore the options while we are able to do so without compromise, so that we can make an investment in a future of self-sufficiency, rather than a continuing series of stopgap measures that are dictated by the urgency of crisis conditions.

NATURAL GAS

Natural gas is the fuel of choice for a wide range of uses in the residential, commercial and industrial sectors of the U.S. economy. It is clean-burning, abundant, and secure from external disruption of supplies. It offers a ready substitute for oil in most applications other than transportation, and development of gas-fueled transportation technologies and infrastructure are under development. Natural gas has numerous feedstock applications; it is making inroads into areas previously reserved for electricity, such as space cooling; and its environmental benefits in combustion applications have led to a resurgence in demand for gas in electricity generation and industrial use.

Although there have been occasional brief shortages in gas supplies for particular regions over the past 10 to 15 years, availability of this energy source generally has been quite stable. The major pipeline network covers most states in the U.S., extensive distribution networks exist in most of the larger urban areas, and bottled gas products are available to the majority of the rural areas. Gas prices rose sharply in the late 1970s and early 1980s, peaking in 1984. Since that time, prices have dropped steadily in nominal terms and much more sharply in real terms, making gas a very affordable commodity for most current energy uses.

Resource Base

Natural gas initially was discovered when drilling for oil, and it generally was considered to be a waste product and was flared (burned) at the wells to dispose of it. Although there were some localized uses for the gas, it was not until the development of relatively extensive pipeline systems for transportation over long distances that gas began to develop a market value that would justify its production separate from oil. This "associated gas," which is produced from oil wells, remains a major share of gross annual gas withdrawals from wells in the U.S. In 1986, from gross withdrawals of about 19 trillion cubic feet (Tcf) of gas, about 5 Tcf was produced from oil wells and the remaining 14 Tcf from gas wells. (To place these numbers in the context of other fossil fuels, one Tcf of natural gas equates in heat content to about 178 million barrels of oil or to about 47 million short tons of coal.)

Natural gas reserves in the U.S. are somewhat greater than oil reserves, with 1985 estimates of about 193 Tcf of proved reserves, which equates to about 11 years of supply for the U.S. at current levels of usage (Figure 1-8). Proved reserves have been dropping slowly over the past several years as the annual additions to reserves have been falling below annual production levels.[37] Federal government estimates of undiscovered recoverable reserves of natural gas in the U.S., made in 1980, are about 590 Tcf, or about 34 years of U.S. supply at current levels of use.[38] More recent estimates that were made by the Potential Gas Committee, affiliated with the Colorado School of Mines, are somewhat more optimistic at 739 Tcf of undiscovered recoverable reserves. This is enough for another 43 years at current levels of consumption.[39]

It is important to note that all estimates of reserves for mineral resources, both proved and undiscovered, are highly dependent on economic factors. As the market price for natural gas increases, drilling

Figure 1-8

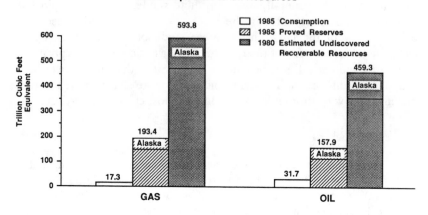

**U.S. Natural Gas Resources
Are Comparable to Oil Resources**

Source: Energy Information Administration

of exploratory and development wells also increases, particularly in areas where previously it was not cost-effective to do so.   Increased drilling expands the base of knowledge of gas reserves, and the price increases change the definition of recoverable reserves by adding those volumes that have become economic to produce at the new price.

About two-thirds of both proved and undiscovered reserves in the U.S. are located onshore in the lower 48 states.   In addition to the reserves noted above, there are estimates that unconventional gas supplies (gas shales, tight gas sands, gas in coal deposits, gas hydrates) could support another 200 years of U.S. consumption at current rates.[40] Of course, this assumes that both the technology is in place to produce these reserves and the price would support their production.

Of the proved reserves, about 33 Tcf are located on Alaska's North Slope and currently are undeliverable to the lower 48 states.   Two proposals have been made, and efforts initiated, to get the Alaskan gas reserves produced and to market.   The first was the Alaskan Natural Gas Transportation System (ANGTS), which was proposed by a consortium of large oil and gas producers.   The project is a pipeline from the Prudhoe Bay area of Alaska, through Canada, and to the northern tier of the U.S. Although some segments of the pipeline have been constructed in Canada and are in use for other purposes, the full project has never proved to be economic.   Price estimates for gas delivered to the U.S. border have been in the range of $15 per thousand cubic feet (Mcf), while the average price for Canadian imports has never exceeded $5 per Mcf and currently is in the $2 range.

The more recent Trans-Alaska Gas System (TAGS) project proposes a gas pipeline to parallel the TAPS oil pipeline from Prudhoe to the port at Valdez, Alaska. At that point, the gas would be liquefied and shipped by tanker to nations on the Pacific rim (Korea, Japan, Taiwan). If this project were to be completed, and energy markets in the lower 48 states would support the costs, it would not be difficult to construct the necessary LNG receiving facilities on the U.S. West Coast and route the tankers there. Price estimates for this gas are in the $5-$6 per Mcf range, and there is greater likelihood that such a project might succeed if it can compete effectively with LNG from other sources, such as Indonesia.

Imports of natural gas into the U.S. are virtually all from Canada and for the past several years have totalled only about 4% of total U.S. supply. Recent trends have shown an increase, however, with more than 6% of total U.S. gas supply coming from imports in 1988. Total proved reserves in Canada and Mexico are almost equivalent to those in the U.S., and their annual production levels are much lower. This situation offers an extension of the secure supplies of natural gas for the U.S. which makes this energy resource much less vulnerable to any of the types of disruptions that have been experienced in the oil market.

Until 1972 the U.S. used half of the world's natural gas. Only a few other nations used it in any significant amounts. Now it supplies one-fifth of the world's commercial energy and 18% of total energy, or almost half of that supplied by oil.[41] Although gas supplies will never approach the international transportability of oil, large low-cost reserves can be put to other uses, such as methanol production; and it is important to understand their distribution. Studies done at the end of 1986 show that of the world's proven natural gas reserves (about 3,600 Tcf), more than 40% lie in the Soviet Union, and another 25% in the Middle East.[42]

End Use

Gas currently accounts for about one quarter of U.S. energy consumption, about equal to coal. The U.S. uses about one-quarter of the gas consumed each year throughout the world, even though it has just 5.7% of the reserves (see Figure 1-9). Its heaviest use is in the industrial sector, which consumed almost 6 Tcf in 1987, followed by residential use at more than 4 Tcf. Electric utility use is growing, at about 3 Tcf in 1987; and the commercial sector used about 2.5 Tcf. Another 1.5 Tcf was used to fuel the natural gas production and delivery system.[43]

Figure 1-9

1985 Regional Market Shares Of Natural Gas
Production And Reserves
— *PERCENT* —

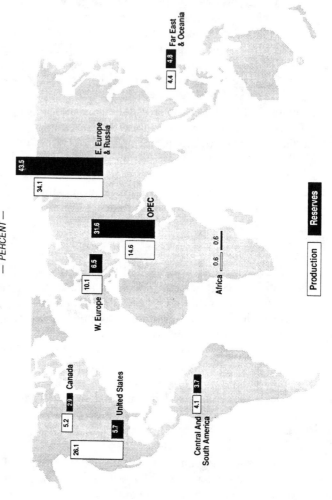

Source: U.S. Energy Information Administration, October 1987

Gas competes at the margin with oil for steam generation, primarily in the industrial and electric utilities sectors. A large number of power plants and industrial facilities are dual-fueled with oil or gas, and which they burn is dependent on the short-term price fluctuations of those commodities. When the price of oil was cut in half in 1986, more than one Tcf of gas use in these sectors was switched to fuel oil to gain the benefits of lower prices. When oil prices recovered somewhat and stabilized in 1987, and natural gas prices fell by more than 25% for industrial and utilities users, most of these users switched back to gas.[44]

This tie to oil prices also works the other way. In the event of another supply disruption, where the price of oil may rise sharply, it is logical to assume that the price of natural gas also would rise. Further, such an increase in prices would not be limited to the sectors mentioned above. The residential and commercial sectors also would be affected, since the market would tighten and wellhead prices would increase until supply and demand came into balance. The key point in this discussion is that although the level of domestic production ensures security of supplies, it does not ensure protection from price increases due to competition at the margin for scarce energy resources.

Natural gas demand in the U.S. dropped steadily from 1973 until 1986, primarily as a function of increasing gas prices. In 1973, delivered gas prices averaged about $0.73 per Mcf; however, by 1984, this average had increased to $4.85 per Mcf in nominal dollars. By the end of 1988, this price had dropped by about 20%, with much sharper decreases in the utilities and industrial sectors. This has led to increasing demand for gas in both 1987 and 1988.[45] Adding to this impetus are the recent environmental pressures in the areas of acid rain and global warming. A major part of new utility construction programs is for gas-fired cogeneration and combined cycle plants, taking advantage of low capital costs, low fuel costs and no problems under strict emissions controls for SO_2, NO_x or CO_2. These changes will lead to a steady increase in gas use in this sector.

There is substantial potential for expanded natural gas use in the transportation sector, as well. As was discussed in the earlier section on oil, development of alternatives to petroleum products, such as gasoline, are critical to our energy security. Further, the same environmental concerns that are creating pressures for cleaner stationary combustion facilities weigh heavily on the automobile and its internal combustion engine. Compressed natural gas (CNG) vehicles offer a much cleaner alternative to gasoline and diesel powered vehicles, although there is some loss in power and flexibility of use.

Fuel form will be the greatest barrier to gas penetration of the automobile fuel market. While 30,000 vehicles are now operating on natural gas in the U.S. and 375,000 currently exist worldwide,[46] the switch to gas as a transportation fuel faces formidable obstacles in widespread implementation. Problems with the current capital stock of automobiles, trucks and buses include a loss of space for the fuel tanks and a significant weight penalty. The cost of vehicle conversion typically is in the $1,000 to $1,500 range, depending on the number of storage tanks. The operating and maintenance costs of the compression equipment for refueling will add another $0.10 to $0.20 per gallon equivalent of gasoline.[47]

It would be feasible to phase-in such a transition by initially modifying fleet vehicles, such as taxis, buses and delivery vehicles. This was done in Tokyo, and the cleaner emissions have made a very noticeable difference in that urban environment. A similar market exists in the U.S., particularly in those regions where smog problems are a direct result of automobile exhausts. The South Coast Air Quality Management District, which oversees the Los Angeles area, already has major test programs for vehicles in place; and several other cities have initiated such tests.

The more difficult problem for widespread use of CNG is the refueling infrastructure. The CNG refueling process is complex, requiring substantial investment in compressor and other refueling equipment. DOE has estimated, based on Canadian experience, that a fully equipped CNG station would cost about $300,000. They also estimate that in order to save one million barrels of oil per day, it would require 2 Tcf of natural gas to be distributed through 40,000 dedicated stations, at a total cost of about $12 billion. Although the refueling process could be simplified, the technology to do so has not been developed.[48]

Environmental Desirability

Natural gas burns more cleanly than petroleum or coal. For an equivalent energy output, gas combustion produces very low emissions of SO_2 (about one pound per billion Btu as compared to several pounds per million Btu for some coals). Since there is no elemental nitrogen bound-in with the methane in natural gas, as there is in the complex hydrocarbons of oil and coal, NO_x emissions are much lower than gasoline (but higher than methanol) and much easier to control in the combustion process. The lower ratio of carbon to hydrogen in natural gas results in more complete combustion of the carbon and leads to lower levels of carbon monoxide (CO), CO_2 and other hydrocarbons. CO_2

emissions from gas are about half the levels produced by coal. Natural gas also produces lower particulate emissions than middle distillate fuels. Furthermore, gas combustion produces virtually none of the solid waste, scrubber sludge or water pollution that is characteristic of coal and oil products.

Economics of Natural Gas Markets

Natural gas regulation has had a somewhat checkered history, and some understanding of that is necessary to comprehend the aberrations in its pricing that have occurred over the past 15 years. Once the pipeline system began to develop, problems with interstate commerce also arose; and the Natural Gas Act of 1938 was passed to regulate interstate pipeline commerce. This regulation was extended to wellhead prices in a 1954 Supreme Court decision. The wellhead price of natural gas was very low until the early 1970s, and the principal costs were in the delivery system. (The average wellhead price for gas in 1973, the year of the oil embargo, was $0.22 per Mcf.)

The regulatory system for interstate gas pipelines was different from other transportation systems in that the pipelines owned the gas that they were transporting. They purchased the gas from the producers at controlled prices, were allowed under Federal regulations to add costs and a "fair rate of return" for the transportation (this rate of return often averaged 15 to 18%), and then sold the gas to local distribution companies (LDCs) that were regulated by the state governments (see Figure 1-10). Producers generally signed long-term contracts with the pipelines to ensure that they had a steady market for their production. Contracts of 20 years' duration were not uncommon through the 1960s, and they generally had some level of "take-or-pay" requirement for the protection of the producer.

This system worked fine as long as prices generally were low and there was not a great differential between what producers received from selling their gas into the unregulated state (intrastate) markets and the Federally regulated inter-state market. Following the 1973 embargo, such a differential appeared; and producers began to sell into the state markets in order to get higher prices. Shortages began to appear on the major pipeline systems, and the problem peaked in the cold winter of 1976-77 when several major cities narrowly missed being forced to cut-off supplies to residential customers because the interstate pipelines could not purchase the gas required to serve them.

The Natural Gas Policy Act (NGPA) was passed in 1978 to bring these two markets together and provide incentives for increased

Figure 1-10

Federal and State Regulation of Natural Gas

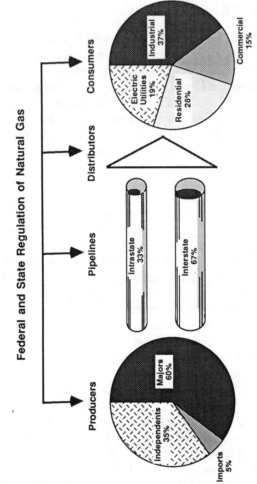

Note: Some pipelines deliver gas directly to industrial and electric utility customers without going through distributors.

Source: U.S. Department of Energy - *Energy Security* Study, 1987

production of new sources of natural gas. The Power Plant and Industrial Fuel Use Act (FUA) also was passed during this period to restrict utility and industrial use of gas under invalid assumptions that the nation was physically running-out of gas and the amount that was left had to be reserved for priority uses, such as residential heating. The result was a maze of regulations that controlled gas prices based on its vintage (when it was discovered) and the depth of the wells, among other criteria. Prices for "old gas" were held to earlier low levels; newer vintages were allowed to move on a path to deregulation that was tied to oil prices. (When the bill was passed, oil prices were projected to be about $15 per barrel when phased deregulation of natural gas was to be complete; so this dollar figure was incorporated into the legislation as the deregulated price.) The most expensive gas to find, such as wells drilled deeper than 15,000 feet, was fully deregulated.

This regulatory scheme, combined with a period of rapidly rising oil prices, gave incentives for high-cost production while giving only disincentives to produce the low-cost gas that was readily available. Gas prices rose sharply while gas use declined. Many pipelines were left with take-or-pay contracts for very high-cost gas while their markets had shrunk, with LDCs unwilling to buy such expensive gas. They still, however, would not transport low-cost gas for others on their pipelines since that might displace the high-cost gas that they owned and were trying to sell to the same customers. This situation resulted in an excess of available supplies for which there was no market, and this "gas bubble" has remained in the market for several years.

In the early 1980s, as oil prices fell, many customers began switching to oil; and pipelines, stuck with take or pay commitments for high-cost gas, started cutting back on purchases of low-cost gas, causing the overall increase in average gas prices to consumers. Customers who were obliged to buy minimum purchases at these inflated prices, appealed to the Federal Energy Regulatory Commission. In the years that followed, FERC issued several orders in an attempt to rectify the natural gas morass but several problems remain. "Old gas" ceiling prices were "devintaged" in an attempt to bring prices more in line with market reality. In addition, FERC declared the minimum purchase provisions between pipelines and customers void. As a result, by the end of 1986 almost half of all gas coming to market was being sold directly from producers to distributors and end-users with interstate pipelines merely transporting the gas for a fee. This share of direct producer to consumer gas sales reflects an increase from a level of only 3% in 1982.[49]

Although the system is not perfect, it is much improved over the first half of this decade. Natural gas prices peaked in 1984 and have been dropping steadily since then. Price competition has resulted in spot market prices for gas that have stayed at about $1.50 per Mcf for an extended period. Full deregulation would be more efficient in the near-term, but this will occur over time as the older vintages of gas are depleted.

Chapter One: Summary

In 1987, the United States economy consumed 76.8 Quads of energy. Of that amount, 89% was produced from fossil fuels--oil, natural gas, and coal. Oil constitutes nearly half of fossil fuel use. These statistics are sobering, given recent experiences with unpleasant economic and environmental consequences of our heavy reliance on oil. The U.S. has experienced oil price shocks that first tripled and then doubled the price of oil, fueling the highest level of inflation experienced in this century. Increasing attention has focused on world environmental problems such as air quality, acid rain, and global warming, all caused by the combustion of fossil fuels. It seems clear, then, that the need for alternatives to fossil fuels will continue to grow.

Oil

Free world energy use has grown at an annual compound rate of over 3%, more than tripling since 1950, and petroleum has accounted for an increasing share of energy use. Today, the world gets nearly half of its energy from oil. In the U.S., 63.5% of total oil use occurs in the transportation sector, where 98% of vehicles depend on petroleum-based fuels. Consumption from this sector alone exceeds the amount of oil the U.S. produces domestically. A remedy for our heavy reliance on foreign oil, then, clearly will require alternate transportation fuels.

As demand for oil has grown, the abundance of low-cost domestic reserves has peaked, and domestic production has declined. This trend has accelerated as a result of low oil prices in recent years. Today, the U.S. holds only 5% of the world's proven economically recoverable reserves. Yet, we are still the world's largest consumer of oil. Most analysts agree that sometime in the 1990s, the U.S. will import more than half of the oil it consumes.

The U.S. is familiar with the potential consequences of heavy reliance on imported oil. In 1973, the OPEC oil embargo caused prices to skyrocket. A tight oil market was hit with a relatively small disruption, and the price of oil tripled overnight. In the 1950s and 1960s, U.S. industries found petroleum products to be an ideal and cheap energy source. Many industrial and commercial users and home-owners converted to oil during that period; and once an initial investment has been made in energy capital stock, it is a long-term proposition to replace it. It appears that the experience of the 1970s was not enough to significantly alter our preference for oil as an energy source.

With oil still playing such a major role in our energy mix, the implications of our growing dependence on foreign imports of this commodity are clear. The most serious of these is our increasing vulnerability to the energy and economic effects of another oil supply disruption, since almost two-thirds of the world's recoverable reserves are located in the Persian Gulf. Recent events, such as the Iran-Iraq war, underscore the tenuous nature of the supply lines from that region.

The economic consequences of dependence on oil imports are serious, both when prices are high and when they are low. Low oil prices spark demand, causing imports to increase further. Increasing oil imports have accounted for a significant share of the U.S. trade deficit. At home, lower prices have discouraged drives toward alternate energy sources, and slowed domestic oil exploration. Reduced exploration activity results in structural changes to the U.S. energy market, such as the exit of highly-trained engineers, which can delay recoveries in exploration activity when prices go back up. Rising oil prices have touched off severe inflation, and a concurrent increase in interest rates. The end result has been recession, with an estimated cost to industrialized nations of $500 billion in lost economic growth projected as the ultimate effect of the 1973-74 oil price hikes.

Natural Gas

The fuel of choice for a wide range of uses is natural gas. It is clean-burning, abundant, and secure from external supply disruptions. It is easily substituted for oil in most applications other than in the transportation sector, where technology and infrastructure development is underway. U.S. reserves are somewhat greater than are oil reserves, at about 11 years of supply at current use rates. Imports of natural gas into the U.S. are virtually all from Canada, and for the last several years have totalled about 4% of total U.S. supply. Proved reserves in the U.S. and Mexico are almost equivalent to those in the U.S., which makes this energy resource much less vulnerable to any of the types of disruptions that have been experienced in the oil market.

The potential for expanded natural gas use in the transportation sector resides in perfecting technologies such as CNG vehicles. However, there are formidable obstacles to widespread implementation of natural gas in the transportation sector. These include problems with the current capital stock of automobiles, trucks, and buses, and the high costs of compression equipment necessary for refueling.

Chapter One: Vital Statistics

Oil

Free world energy use has tripled since 1950.

Almost half of the world's energy comes from oil.

Over 40% of U.S. energy comes from oil.

- Two-thirds of U.S. oil consumption is for transportation
- U.S. transportation consumes more oil than the U.S. produces.
- Early in the 1990s, the U.S. will import more than half the oil it consumes.

Two-thirds of the world's recoverable oil supplies are in the Persian Gulf. Largest include:

- Saudi Arabia   27%
- Kuwait   15%
- Iran   8%
- Iraq   8%

Saudi Arabia has a reserves to production (R/P) ratio of 95; Kuwait 182.

U.S. R/P ratio is 8.

U.S. oil prices over the 1980s have been as high as $44 per barrel and as low as $8 to $10 per barrel (1985 $).

The average U.S. oil well produces 14 barrels per day.

The average Saudi Arabian oil well produces 7,000 barrels per day.

From 1976 to 1986, U.S. consumers paid $640 billion for imported oil, an amount equivalent to the U.S. trade deficit for the same period.

Natural Gas

The U.S. has about 193 Tcf of proved gas reserves, or 11 years of supply. Another 739 Tcf (43 years) of undiscovered recoverable reserves are estimated to exist.

Gas prices in the 1980s have ranged from less than $1 per Mcf to more than $10 per Mcf.

Gas provides about one-quarter of U.S. energy consumption.

Compressed natural gas (CNG) vehicles would cost $1,000 to $1,500 per vehicle conversion.

CNG refueling capacity to offset 1 MMB/D of oil would cost about $12 billion.

Chapter One: End Notes

[1]United States Department of Energy, *Energy Security: A Report to the President of the United States,* (Washington, D.C., 1987) p. 12.

[2]Daniel Deudney and Christopher Flavin, *Renewable Energy: The Power to Choose* (New York: W.W. Norton, 1983), p. 10.

[3]*Energy Security*, p. 12.

[4]*Monthly Energy Review, op. cit.,* pp. 9, 29.

[5]*Ibid.*

[6]Larsen, Robert P. and D.J. Santini, "Rationale for Converting the U.S. Transportation System to Methanol Fuel," May 1986, n.p.

[7]Robert Stobaugh and Daniel Yergin, *Energy Future: A Report of the Harvard Business School* (New York: Random House, 1979) p. 18.

[8]*Energy Security,* p. 17.

[9]"The Next Oil Crisis," *Commonweal,* March 3, 1987, p. 132.

[10]*Energy Security,* p. 3.

[11]U.S. Department of Energy, *Monthly Energy Review,* August 1988, p. 37.

[12]*Energy Security,* p. 8.

[13]James Tanner, "Panic After Any Gulf Oil Cutoff Wouldn't Last Long, Analysts Say," *The Wall Street Journal* (July 23, 1987), p. 26.

[14]*Energy Security,* p. 17

[15]*Ibid.,* p. 18

[16]*Ibid.,* p. 27

[17]*Energy Security,* p. 15.

[18]*Monthly Energy Review, op. cit.,* p. 36.

[19]*Ibid.,* p. 93.

[20]Barry, John M., "Congress Pushes Homegrown Energy Again," *Business Month,* November 1987, p. 70.

[21]*Energy Security,* p. 28.

[22]*Ibid.*

[23]Tanner, James, "Cartel's Comeback? By Early 90's OPEC May Dominate Oil Market," *Wall Street Journal,* November 21, 1987, p. 15.

[24]*Energy Security,* p. 6.

[25]Michael R. Fox, "Future Energy Supplies," *Vital Speeches,* (March 30, 1985) p. 554.

[26]Solomon, Caleb, p. 8.

[27]*Energy Security,* p. 6.

[28]*Ibid.*

[29]*Monthly Energy Review, op. cit.,* p. 36., p. 64.

[30]Tanner, *op. cit.,* p. 23.

[31]Robert L. Hirsch, "Impending United States Energy Crisis," *Science* (March 20, 1987), p. 1470.

[32]Klein, *The Houston Chronicle,* p. 27.

[33]Klein, *op. cit.*

[34]*Monthly Energy Review,* p. 37.

[35]U.S. Department of Energy "Assessment of Costs and Benefits of Alternate Fuels in the Transportation Sector," January 1988, p. ii.

[36]Stobaugh and Yergin, *op. cit.,* p. 40.

[37]U.S. Department of Energy/Energy Information Administration, *Natural Gas Annual 1986, Volume 1,* Washington: USGPO, October 1987, p. 12.

[38]*Energy Security, op. cit.,* p. 116.

[39]"Estimate Cut for U.S. Potential Gas Resources," *Oil & Gas Journal,* April 27, 1987, p. 105.

[40]*Energy Security, op. cit.*

[41]Deudney, Flavin, p. 17.

[42]*Ibid.,* p. 117.

[43]*Monthly Energy Review, op. cit.,* p. 58.

[44]*Ibid.,* pp. 58, 105.

[45]*Ibid.*

[46]*Energy Security, op. cit.,* p. 120.

[47]U.S. Department of Energy, *Assessment of Costs and Benefits of Flexible and Alternative Fuel Use in the U.S. Transportation Sector,* Washington: DOE/PE, pp. E-4, E-5.

[48]*Ibid.* p. C-4.

[49]*Energy Security, op. cit.,* p. 124.

CHAPTER TWO

COAL USE AND THE ENVIRONMENTAL IMPACTS OF FOSSIL FUEL COMBUSTION

Time figures prominently in assessing the risks of human-induced changes in the earth's chemistry. Given sufficient time to adjust, the earth and its organisms exhibit remarkable resilience in the face of change. Yet during that period of adjustment--whether to higher global temperatures, a blanket of air pollutants, acidic rain, or toxic chemicals--much suffering, economic loss, and social disruption may occur. As a dominant agent of change, humanity must confront the adverse consequences of that change, and protect this and future generations from them.

-Sandra Postel

The principal problem with coal as an energy source is environmental damage that results from its combustion. The use of coal to generate heat and electricity results in emissions of gases that cause acid rain, negatively affect air quality in other ways, and may contribute to global climate change and its potential impacts. Combustion of other fossil fuels produces many of the same emissions, but not to the degree coal combustion does. This chapter begins with a discussion of coal use in the United States, and then reviews the major environmental concerns associated with fossil fuel combustion in general.

COAL USE

Coal is by far our most abundant fossil fuel. Demonstrated reserves in the U.S. alone approach 500 billion tons, an amount that would supply our domestic needs for several hundred years at current consumption levels.[1] Moreover, worldwide recoverable reserves hold about three times as much energy as oil and gas reserves combined.[2] Four countries--the Soviet Union, the United States, the People's Republic of China and Australia--contain about 90% of the world's total coal resources and 60% of the proved, economically recoverable reserves.[3]

For much of U.S. history, coal was our primary fuel source. It has been mined here commercially for over 240 years, and coal provided the energy that fueled the industrial expansion of the late 19th century. By the beginning of the early 20th century coal began to lose dominance as the infrastructure of the oil industry developed. The expansion of gas and oil pipelines led to the replacement of many coal fired locomotives, industrial boilers, and electricity generators with those that burned oil and gas. In addition, the new availability of cheaper Middle Eastern oil and the emergence of concerns about air pollution accelerated the trend away from coal use. While coal provided nearly 90% of all U.S. energy from mineral fuels in 1910, it accounted for less than 20% of that total in 1970.[4]

In 1977, with the advent of the second energy crisis, President Carter called for an 80% increase in coal consumption as part of his National Energy Plan. The price shocks of the 1970s encouraged the return of coal to uses previously dominated by petroleum. In the last two decades there has been an increase in coal use for electricity generation, with some older power plants being converted back to coal and a greater share of new electric power plants designed to burn coal. Furthermore, coal is being used more frequently in other industrial applications, such as a fuel for industrial boilers and in cement and lime kilns. As a result, since 1974, domestic consumption of coal has increased by almost 25% (Figure 2-1).[5] Even with these increases, however, coal remains an energy resource for which its potential far outweighs its use. Today, although coal accounts for 90% of total U.S. energy reserves, it currently provides only about 23% of the energy used in the United States.[6] The Department of Energy, however, is projecting that U.S. coal use will grow 25% in the ten-year period 1985-1995; and that worldwide consumption of coal is likely to expand by more than one-third in that same period.[7]

While coal is relatively low-cost and abundant, it has its disadvantages. Coal's bulk and lower heat content per unit of volume, characteristics that restricted it to a secondary role in the past, will continue to impede its return to dominance in the future. It takes only four barrels of oil to equal the heat content of one ton of coal; and the liquid form of hydrocarbons is much less difficult to produce, process, transport and store. Finding a produceable coal deposit is not difficult; but mining it usually is, particularly for labor-intensive efforts such as those required in eastern deep mines where productivity is an ongoing concern. Coal processing is a limited and relatively unsophisticated stage of the production cycle, although more and more research dollars are being put into technologies for coal cleaning--to remove in advance the sulfur and other minerals that become pollutants when coal is burned.

Figure 2-1

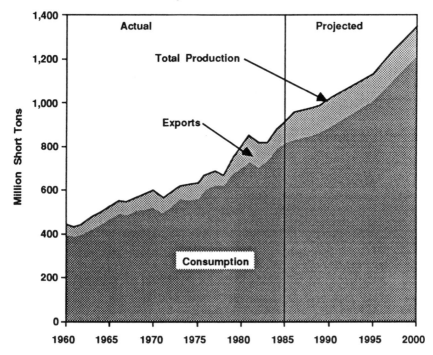

Actual and Projected U.S. Coal Production

Source: U.S. Department of Energy - *Energy Security* Study, 1987

Coal transportation is a major problem with expanded coal use and often is the most expensive to solve. Coal transportation normally is about one-third of the total cost of delivered coal, but those costs can exceed two-thirds, depending on the quality of the coal and the distance to market. The U.S. is the highest-cost coal supplier on the world market, and it is generally conceded that the principal reason for these high costs (and the accompanying loss of world market share) is the cost of inland transportation to sea ports for shipment.

Most coal is shipped long distances by rail (more than 60%) and barge, followed by truck and mechanical conveyors for short distances.[8]

Numerous efforts have been made over the years to construct coal slurry pipelines to transport a coal and water mixture less expensively over long distances; however, these efforts consistently have been curtailed by the inability to get rights-of-way for the pipelines, primarily from those lands held by the railroads. Although some progress has been made in the past few years to improve rail efficiency and stabilize costs to shippers, coal transportation will continue to be a hurdle that must be overcome for expansion of U.S. coal use.

The problems inherent in coal handling and storage at the utility or industrial facility will continue to inhibit expansion of industrial coal use. Industries located in urban areas are nearly precluded from implementing new coal applications, simply because of the space requirements for coal storage and handling equipment. Further, coal use requires higher initial capital investment; and this generally will limit new coal applications to those areas where large capacities and high utilization rates can compensate through early amortization of these costs.

The full benefits of burning coal are best gained in the economies of large, high-volume applications, that can be appropriately situated for efficient transportation and handling. This leads to electric utility use as the principal application, and almost 85% of the coal consumed in the U.S. in 1987 was burned by electric utilities. However, the market for baseload electricity generation has changed dramatically in the past few years; and the implications of these changes offer an uncertain future at best for continued steady growth of coal-fired electric generation capacity. The two key factors in these market changes that are affecting coal use deserve some additional discussion. These are:

o   A prevailing view among both utilities and the financial (lending) community that the market will not support investment in large, capital intensive, power plant construction projects that will take several years to complete; and

o   The potential for Clean Air Act revisions that would both add substantial retrofit costs to older coal-fired plants not subject to the original Act and tighten the requirements for newer plants.

Economic Environment for Electric Utilities

Historically, electric utilities have been able to finance and construct baseload power plants of increasing size, taking advantage of economies of scale to keep consumer electricity rates from rising too rapidly. Electric utility markets have gone through a dramatic period of transition over the past decade, and these changes are not yet complete. One investment analyst characterized those years as a financial crisis for U.S. utilities that was ". . . at least as bad as the Great Depression." Electricity load growth, which had averaged about 7% per year for several decades, fell to an average of only about 2.5% from 1973 to 1982. Based on earlier expectations of continued high growth, many utilities had made major commitments to new construction. As expected loads failed to materialize, these utilities found themselves facing substantial amounts of excess capacity. More than 200 plants, both coal (90 plants) and nuclear, were deferred or cancelled. Despite this response, the industry still reached an average reserve margin of almost 40% in the early 1980s-- double the level that generally is considered prudent.

These difficulties were compounded by the effects of extremely high inflation, construction delays that caused substantial cost growth in virtually all projects, and pricing regulations that did not allow utilities to recoup construction costs until new plants were on-line and generating power. As a result, utilities, which traditionally had assumed they operated in the low-risk environment of assured markets and favorable regulatory treatment, suddenly found themselves confronted with extreme risks to their financial well-being.

Utility problems with completing plants on-time translated into sharp increases in consumer prices for electricity in several regions of the country. This led to a strong consumer movement against electricity rate increases, which brought increasing pressure on state public utility commissions (PUCs), leading most of them to become very sensitive to consumer concerns. In the current environment, many PUCs have developed policies that encourage utilities to meet new load growth through demand management and other conservation techniques, thereby avoiding new construction and the rate increases that would accompany it.

The lesson that utility executives and their boards have learned from the period just described has been risk avoidance, which has tended to translate into strong opposition to the financial risks associated with any new baseload plant construction. Even those utilities that have identified substantial requirements for new capacity are using every available option, short of new baseload construction, to meet growth in demand. One top utility executive, in a recent speech, stated:

Much has been written about the fracturing of the long-standing regulatory compact. The concept of utility regulation historically mirrored the scales of justice--a balance between consumer and

*investor interests in a monopoly setting. Economic events of the
past 15 years have caused regulators to tip the scales sharply in
favor of short-run consumer interests.*[9]

Some companies have even announced that they will *not construct*
baseload generating capacity in the future. Smaller increments of power
are available from a variety of sources, and utilities in several states are
now using competitive solicitations to contract for long-term power
supplies. A class of independent power producers (IPPs) has developed
that uses the cogeneration advantages of the Public Utilities Regulatory
Policy Act (PURPA) to gain financing and access to the market. This
group includes small and medium sized entrepreneurs and even some
subsidiaries of major utilities that are operating outside their normal
service areas.

In summary, the current economic environment for utilities, and
the environment that is projected for the next several years, is one that is
biased strongly against any large commitment of capital for plant
construction. As one top utility executive expressed, the next several
years will witness a ". . . strategy of gradual incrementalism." In his view,
most utilities will continue into the 1990s using low-capital cost
approaches to supply as much as 20% of a utility's total capacity,
including: power imports; contracts with IPPs; cogeneration; demand-
side management; and combustion turbines. It is a widely-shared view
that the bias against large utility capital outlays is so great that the
financial community will support very few proposals for large coal-fired
generation--and none for nuclear generation.

Environmental Problems and their Control

Emissions from Coal Combustion

The principal problem with coal that has emerged over the past
two decades is its impact on the environment. Although there are similar
environmental impacts from combustion of oil and natural gas, the
problems associated with coal-burning are both more severe and more
expensive to control.

Coal-fired utilities are the primary source of sulfur dioxide (SO_2)
emissions, the key precursor to acid rain. The Federal government
estimates that sulfur emissions in 1980 totaled 23.6 million metric tons
(mmt). Of that amount, 15.9 mmt came from utility power generation;

and 14.7 mmt was from coal-fired utilities.[10] Emissions of SO_2 have been dropping, albeit slowly, since the early 1970s and the implementation of the Clean Air Act. By the end of 1986, total utility emissions of SO_2 had dropped to about 14.4 mmt.

Coal-burning also produces higher levels of nitrogen oxides (NO_x) than combustion of either oil or gas. NO_x emissions are precursors to both the nitric acid component of acid deposition and the production of tropospheric (low-level) ozone. The Federal government estimates that NO_x emissions in the U.S. in 1980 totalled 19.4 mmt, with 5.8 mmt coming from utility combustion. However, the largest single contributor to NO_x emissions is not power generation, but the internal combustion engine--cars and trucks in the transportation sector, which produced 7.2 mmt of NO_x in 1980. Total levels of NO_x emissions have remained relatively flat in the last several years, with 1986 levels of 19.1 mmt. Utility emissions have increased in that period, however, to 6.3 mmt.[11]

Because of its high ratio of carbon to hydrogen, coal combustion produces higher levels of carbon dioxide (CO_2) than either oil or gas. Since CO_2 is produced in all fossil fuel combustion and many other natural processes, such as decaying vegetation and breathing, it was not until recently identified as a problem. Advances in atmospheric modelling have led scientists to almost a consensus view that increasing CO_2 concentrations in the atmosphere are leading to a global warming effect that could have severe environmental impacts in the coming century.

Controlling Emissions

The options for abatement of sulfur emissions from coal combustion fall generally into two categories: fuel switching, and technology solutions. Fuel switching takes advantage of the fact that the sulfur content of coal varies greatly, from very low levels of about 0.2% by weight to more than 5%. Low sulfur coal is common in the West, in states like Wyoming and Montana. It constitutes more than half of our coal reserves by weight, but generally is a lower-rank, subbituminous coal that has about 30% less energy per ton than Eastern coal. It usually is strip-mined, which is a much less labor-intensive operation than underground techniques.[12] The coals with the highest sulfur content are found in the Northern Appalachians and in the Midwest. Eastern high sulfur coal, which generally has higher heat content per ton than Western coal, is most often mined deep underground in a very labor-intensive environment. The tradeoffs for fuel-switching focus on the delivered costs of low-sulfur coal over the useful life of the plant in comparison to the costs of adding scrubbers.

In the area of technology solutions to SO_2 reductions, a variety of advanced approaches are being examined. The most common technology, however, is wet scrubbing of the flue gas, using a form of water and limestone injection that is sprayed directly into the flue gas stream. Although this is the most used technology, it has several drawbacks. The greatest problem is the high volume of waste that is generated in the process. Advanced technologies for cleaning sulfur from the flue gas are focused on dry processes.

Reduction of NO_x emissions presents a different set of technical problems as these pollutants are generated in two ways. "Fuel NO_x" comes from nitrogen compounds that are contained in the coal or oil (there is virtually no elemental nitrogen in natural gas). "Combustion NO_x" is generated due to an excess of air during high-temperature combustion. Given the conditions that cause combustion NO_x, substantial reductions can be achieved through modifications to the combustion process. These include low-NO_x burners, lower temperature combustion, multi-stage combustion, air recirculation and other controls on combustion patterns. In the more advanced applications, these modifications can achieve about a 50% reduction in NO_x emissions.

To achieve higher levels of NO_x emissions reduction, more expensive technologies may be required, such as selective catalytic reduction (SCR) reactors placed in the flue-gas stream. SCR can achieve reductions of up to 95% of NO_x emissions, but the installations are complex and the costs can be high depending on the composition of the catalyst. The majority of experience with SCR resides in Japan and West Germany, where emissions controls historically have been much more stringent than those in the U.S. Currently there are no major U.S. utility installations of SCR; however, new air quality regulations in the Los Angeles area of California will require SCR on gas-fired utility boilers and turbines in the near future.

Retrofit technologies, such as SO_2 scrubbers and SCR for NO_x abatement, can be quite expensive to install and to operate. In many cases, such retrofits on older plants will not be cost-effective and will result in their decommissioning. New plants will require combustion technologies that can meet current and future emissions limits at a reasonable cost, and the Federal government has committed $2.5 billion to a Clean Coal Technology (CCT) R&D program to develop and demonstrate such technologies. This program pays up to 50% of the cost of CCT demonstrations and will have awarded 60% of its grant monies by the end of 1989. In addition to advanced SO_2 and NO_x scrubbing systems for new plants, the CCT program is supporting research on advanced combustion systems, such as both atmospheric and pressurized versions of fluidized bed combustion (FBC). These approaches remove SO_2 and NO_x in the burning process, before it moves into the flue-gas stream.

The technologies identified above are effective. However, they also can be quite costly. In many cases, it is much more cost-effective for a utility to keep older plants and change to a low-sulfur coal to meet SO_2 emissions limits. Fuel switching, however, may not be a universal option for older plants under the current proposals for acid rain legislation. The economies of several states could be greatly affected under a system of strict emissions controls, when there are no limits on fuel switching.

It is easy to see why the acid rain issue results in polarization of different regions of the country. Tradeoffs are necessary when trying to use the lowest-cost option for achieving required emissions limits. In many cases, this would be a fuel-switching option, but the high-sulfur mining interests in affected regions would suffer. A second set of tradeoffs is encountered when asking who pays the increased costs, particularly if a high-cost scrubber option is required. The options range from a simple "polluter pays" approach to some relatively complex combinations of electricity and carbon taxes and incentives. Currently, because they rely heavily on older coal-fired power plants, the states of Missouri, Illinois, Indiana, Ohio and Pennsylvania account for more than 40% of annual utility SO_2 emissions.[13] It is logical to assume that under most acid rain proposals, where the "polluter pays" philosophy is prevalent, these same states would bear a heavy cost burden for emissions reduction.

It is quite difficult to track all of the impacts that may be associated with the many acid rain proposals currently before the Congress. The Congressional Budget Office (CBO), an analytical arm of the U.S. Congress, has categorized the major types of programs being proposed and has analyzed both the aggregate and state-by-state impacts of these proposals. The major variables are: the level of SO_2 and NO_x emissions reductions that are required; how much flexibility in fuel switching is allowed; and how the costs of the proposals are to be distributed. NO_x reductions so far have not received the same level of attention at the national level as have SO_2 reductions. Most of the options are discussed in terms of the levels of SO_2 reduction that are mandated, with the key proposals offering reductions from 1980 levels of 8, 10 and 12 million tons per year. The question of "who pays and how" generally examines whether there will be a strict "polluter pays" provision or if some scrubber subsidies or other financial aid will be provided.

CBO has used as a baseline scenario a "polluter pays" approach with no fuel-switching restrictions. Their analysis, in 1985 dollars, shows that costs per ton of SO_2 abated to achieve reductions of 8, 10 and 12 million tons per year, would be $270, $360 and $779 per ton, respectively. If restrictions are placed on fuel-switching, forcing increased scrubber installation rather than substitutions of low-sulfur coal, CBO estimates that these costs would increase to $306 and $528 per ton of SO?2? for the 8 and 10 million ton per year options, making this the most expensive case.[14]

In the baseline case above, fuel-switching by utilities has economic benefits for them but can have dramatic negative effects on the economies of the affected coal-producing states. The states of Indiana, Illinois, Ohio and Pennsylvania are the heart of the high-sulfur coal producing region; and under a continuation of current policies, their production would remain constant over the next decade. Fully a quarter of their coal production would be lost under an 8 million ton per year SO_2 reduction scenario, and 38 percent would be lost under the 10 million ton per year reduction scenario if fuel-switching were allowed. This latter case translates into a loss of coal mining jobs of almost 22,000 in the four-state region.[15]

It is not the purpose of this section to present a detailed analysis of the impacts of acid rain legislation on the coal industry or the utilities. Rather it is to illustrate the range of difficulties and uncertainties facing proponents of increased coal use over the next decade or more. The economics of coal production and use, combined with the regulatory restrictions in place, have relegated coal to its role in baseload power generation. Even this role has uncertain growth potential, given the current economics of baseload power generation, the pending slate of acid rain legislation, and the uncertainties of some form of legislation on global warming.

The potential for coal to substitute for other fuel forms does not appear to be highly probable in the near future. Current technologies for converting hydrocarbons from coal into liquids or gases are not yet economically competitive, and improvement in the economics of these processes is partially dependent on a sharp increase in the price of oil. Therefore, the future of coal use over the next decade or so appears to be constrained to the markets outlined above; and some potential exists for demand to fall as well as rise.

ENVIRONMENTAL IMPACTS OF FOSSIL FUELS

The political and economic costs of our dependence on petroleum can be severe, as has been outlined in the preceding discussion. The consequences of our actions are clear--we had experience with many of them in the oil supply disruptions of the 1970s. In the area of environmental impacts of fossil fuel use, however, neither the causes nor the effects are quite so clear. In several of our major cities, we can see many of the direct effects of fuel combustion in the layer of smog that will blanket them on hot summer days. However, there are many effects of polluted air that we cannot see; and many more are transported to other areas, often over long distances. Neither can we see the global warming effects that may have their greatest impacts 50 years from now. It is difficult enough to keep the public's attention on events that will occur in a few months or even weeks, and the idea of environmental planning to meet goals that are a half-century in the future is almost beyond our ken. However, these are attitudes that must change if we are to assure ourselves both a strong energy future and the environment in which to enjoy it. Along with the advantages we are offered by technology comes the critical responsibility to use it efficiently and with care.

Six elements comprise 95% of the mass of all living matter on earth: carbon, oxygen, nitrogen, hydrogen, phosphorus and sulfur. Their supply is limited and so our life depends upon the efficiency with which they cycle through the atmosphere. Recently, researchers have learned that human activities have significantly disrupted these cycles, most notably those of carbon, hydrogen and sulfur. Since 1860 the combustion of fossil fuels has released some 185 billion tons of carbon into the atmosphere. However, while in 1860 annual emissions were estimated at 93 million tons, at present carbon emissions exceed five billion tons per year, a fifty-three fold increase (see Figure 2-2).[16] In fact, the bulk of all CO_2 emissions from fossil fuel combustion during the 130 year period from the beginning of the industrial revolution until today has occurred since 1950.

The rapid rise in oil-use that began in the 1950s has driven this trend, adding substantially to carbon releases from coal. A significant factor in this increase, especially in the U.S., has been increased consumption for transportation. The U.S. uses more oil than it produces to fuel the transportation sector of the economy. Transportation sources add over 700 million tons of carbon to the atmosphere annually. The average American car pumps its own weight in carbon into the atmosphere each year.[17] The transportation sector currently is responsible for about one-third of man-made CO_2 emissions in the U.S.[18]

The environmental consequences of fossil fuel combustion fall into three broad categories. These are: effects on ambient air quality; acid precipitation; and contribution to global warming (the "greenhouse effect"). Concern over global warming has achieved prominence only

Figure 2-2

World Carbon Emissions from Fossil Fuels, 1950-84

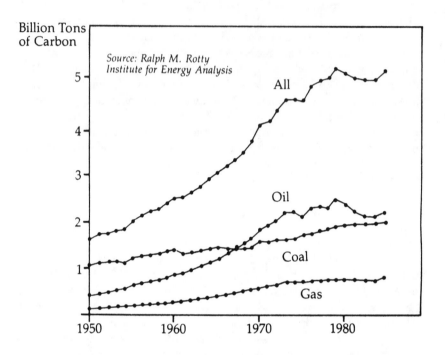

recently, while concerns regarding diminished air quality and acid precipitation are longstanding. The remaining sections of this chapter discuss these environmental concerns, paying particular attention to the extent of these problems, their potential human effects, and the role fossil fuel combustion plays in each.

Air Quality

The U.S. EPA considers several pollutants in defining air quality. In particular, attention has focused on particulates (measured as total suspended particulates or TSP), sulfur dioxide (SO_2), carbon monoxide (CO), nitrogen oxides (NO_x), ground-level ozone (O_3), and lead (Pb). (Figure 2-3 shows the number of persons exposed to concentrations of these pollutants above what is considered safe.) Volatile organic compounds (VOCs) also are important because of the role they play in

Figure 2-3
U.S. Population Exposed to Selected Pollutants
Source: U.S. EPA

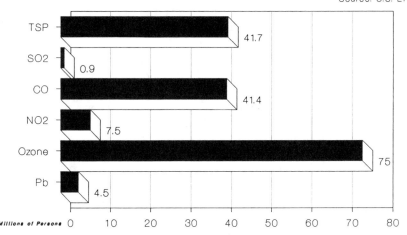

Millions of Persons

Number of persons living in countries with air quality levels above the primary
national ambient air quality standards in 1986 (based on 1980 population data)

the formation of ground-level ozone (as well as their role as precursors to acid precipitation). Over the last 10 years, improvements can be seen in emissions trends and concentrations for all of these pollutants. However, environmental and health threats from these pollutants remain a concern, and fossil fuel combustion is generally the major source for all of them. In general, air quality concerns are greatest in urban areas, especially for those in warmer regions of the country.

The most severe pollution problem, in terms of the number of Americans affected, is tropospheric (ground-level) ozone. In 1986, 75 million people lived in areas where federal air quality standards for ozone levels were exceeded.[19] Ozone, which is beneficial in the stratosphere because it shields the Earth from the most harmful solar radiation, is the major component of urban smog and can have serious human health effects when it occurs at ground level. Ozone acts as a respiratory irritant, aggravating such chronic diseases as asthma, bronchitis, emphysema and heart disease.

Ozone is not emitted directly into the air, but is produced at ground level through complicated chemical reactions involving NO_x, VOCs, and CO. Both VOCs and NO_x are emitted from transportation and industrial sources, including automobile exhaust, electrical power plants, factories, and processes using solvents, such as those occurring at dry cleaners, paint shops, and chemical manufacturing facilities. Carbon monoxide is emitted chiefly from transportation sources, with over two-thirds of total emissions coming from highway vehicles. Sunlight and temperature play important roles in stimulating ozone formation. This means that ozone levels generally are seasonal, with peaks occurring during the warmest months and in the most heavily populated areas.

Nationally, the annual average of daily maximum ozone values decreased by 13 percent from 1979 to 1986. For the 89 largest metropolitan statistical areas (MSAs), the number of days on which the ozone standard was exceeded declined by 38 percent. However, many of these urban areas continue to frequently exceed the maximum standard of 0.12 ppm established under the EPA's National Ambient Air Quality Standards (NAAQS).[20] Thirty-seven cities were far above acceptable levels. Los Angeles, among the worst, suffers over triple the level of ozone considered safe. Concentrations also were high along the Texas gulf coast, the northeast corridor, and in other heavily populated regions.[21]

National emissions of volatile organic compounds (VOCs) are projected to rise from 1985 to 2010, from 27.1 million tons to 28.4 million tons per year. Emissions of VOCs from energy-related activities, principally petroleum production and distribution and transportation sources, are projected to decline between 1985 and 2000. Emissions of VOCs from transportation sources are projected to decline by 35 percent (from 7.2 million tons to 4.7 million tons per year) due to the implementation of emissions standards and the scrappage of older vehicles.

Transportation has historically been the major contributor of carbon monoxide emissions. In 1980, about 67 million tons per year of CO (74 percent of the national total) were emitted from transportation sources. By the year 2000, annual CO emissions from transportation are projected to drop to 37 million tons per year, a 45 percent reduction from the 1980 level. A slight increase is projected after 2000, as vehicle-miles traveled continues to rise.

(For a detailed discussion of NO_x emissions, see the Acid Rain section of this chapter.)

Overall, the transportation sector's contribution of tropospheric ozone precursors has decreased and is projected to continue to do so through 1995 because of existing emissions control regulations and standards. However, ozone problems may persist, mainly in urban areas where there are prolonged periods of meteorological stagnation during warm, sunny seasons, or dense population growth in a hot, arid climate. If annual average temperatures increase under global warming scenarios, those areas that already are projected to experience continuing problems with tropospheric ozone would face even greater difficulties in meeting existing ozone standards.

Particulate matter in the air (measured as TSP) includes dust; dirt; soot; smoke; liquid droplets emitted by factories, power plants, and automobiles; and windblown dust, especially from construction. TSP levels have decreased over the period from 1977 to 1986 by 23 percent, although in 1986 nearly 42 million Americans lived in areas that exceeded federal air quality standards for TSP levels.

Sulfur dioxide (SO_2) levels also have decreased over the last 10 years, by a median rate of 4 percent per year. All of the 89 largest MSAs were below the national ambient air quality standard of 0.03 ppm in 1986. However, most major sources of sulfur oxide emissions are in rural areas, and the direct health effect of these emissions may still be substantial when combined with particulate emissions. The U.S. Office of Technology Assessment estimates that the current mixture of sulfates and particulates in ambient air may cause 50,000 premature deaths in the U.S. each year, approximately 2% of our annual mortality.[22] Sulfur oxide emissions also are of concern because of their role in acid precipitation.

SO_2 results primarily from stationary source combustion of coal and oil, both fuels containing levels of elemental sulfur. A majority of SO_2 emissions are dominated by a relatively small number of large emitters. Nationally, two-thirds of SO_2 emissions are generated by electric utilities, 94 percent of which come from coal-fired power plants. Fifty-three individual plants in 14 states account for one-half of all power plant emissions, and the 200 highest SO_2 emitting plants account for 57 percent of all SO_2 emissions nationally. In 1986, total fossil fuel combustion accounted for 17.2 million metric tons (mmt) of sulfur oxide emissions.[23]

Carbon monoxide, in addition to its role as a precursor to ground-level ozone formation, is a poisonous gas treated individually under NAAQS. The maximum average concentration for an 8-hour period is 9 ppm. Over the last 10 years, the national average for CO concentrations dropped by 32 percent. However, over 41 million people live in areas which exceeded the maximum standard in 1986. Its importance as a pollutant is clear when one considers its role in ground-level ozone formation, discussed earlier.

As a result of concerns about air quality, Congress passed the Clean Air Act (CAA) in 1970. The act originally projected the cleanup of the nation's air by July 1, 1975, when all states were to meet the primary National Ambient Air Quality Standards (NAAQS) for the major air pollutants discussed above. This deadline has been extended repeatedly, however; and today, almost 20 years after the Act was passed, ambient air quality standards for ozone and carbon monoxide have not yet been met in numerous regions of the country. In August of 1988, EPA found that 68 cities did not meet ozone standards and 59 did not meet carbon monoxide standards. Congress currently is poised to renew the Clean Air Act, but legislative solutions clearly have not been entirely successful in achieving desired reductions in pollution levels.

Under the law, EPA can impose stiff sanctions on states that have not met standards under the Act, including cutting off Federal highway funds and curtailing economic development. EPA has been reluctant, however, to take such steps, partly because they would require reductions in emissions from smaller sources and the need to impose tighter standards on automobile emissions. Simply put, the easiest pollution problems to solve technologically have, to a great extent, been addressed. Further progress will be difficult to achieve using conventional approaches, since a myriad of smaller sources pose substantial technological challenges; and it would require both great expense and a tremendous regulatory apparatus to address these effectively. Thus, progress toward further reductions in local air pollution is at something of an impasse.

One current example illustrates this point. Attempts to achieve ozone and carbon monoxide reductions have included efforts to reduce the quantity of gasoline fumes that escape during refueling. The Environmental Protection Agency estimates that nearly 10 pounds of gasoline fumes are emitted into the atmosphere for every 1,000 gallons of fuel pumped into cars.[24] The issue has prompted a fierce and protracted battle among oil companies, service station owners, auto manufacturers, and other interested parties. Oil companies and service station owners are fighting to avoid regulations that would require the installation of pumping systems and underground tanks that would capture the fumes. Auto manufacturers are vigorously opposing an alternative approach that would require the installation of "on-board canisters," as an additional component of existing pollution control equipment.

Rep. John D. Dingell of Michigan, the powerful Chairman of the House Energy and Commerce Committee, has nearly single-handedly held up enactment of an EPA proposal that would require auto makers to install the on-board canisters. Under the EPA proposal, the devices would be required on all vehicles in the 1991 model year.[25] The outcome of this battle, already many years old, remains in doubt; but it illustrates the difficulties "command and control" approaches to pollution problems have faced as competing interests fight to avoid the substantial environmental costs of burning traditional fuels. Further, it seems clear that whatever the ultimate approach, effective enforcement of the plan will place a substantial regulatory burden on already stressed federal and state agencies.

Acid Rain

A second major environmental consequence of fossil fuel combustion is acid rain. Acid rain occurs when sulfur dioxide (SO_2) and nitrogen oxides (NO_x), emitted when coal or other fossil fuels are burned, are transformed through a series of chemical reactions in the atmosphere to produce sulfuric acid (H_2SO_4) and nitric acid (HNO_3) gas, both of which can be deposited either in dry form or dissolved in earthbound rainwater. Acid rain visibly erodes sculptures and damages crops. Beyond these obvious effects, it can have deleterious effects on large ecosystems, including entire forests and watersheds. In lakes and streams it can damage fish populations and catalyze reactions of toxic metals deposited on lake beds, resulting in potentially serious human health effects.

Over 95% of the sulfur emissions produced in the U.S. are from stationary fuel combustion sources, with 60% coming from coal burning electric plants.[26] For every two tons of sulfur dioxide emitted, one ton of sulfur is added to the atmosphere. Presently, man-made emissions of sulfur equal those from all natural sources combined.[27] Nitrogen oxide emissions are split almost evenly between stationary and mobile sources, with 47% from fixed installations (28% from power plants) and 44% from transportation activity.[28]

Sulfur emissions are dropping steadily, as we have discussed, and are projected to continue to decline as older coal-fired power plants, which are generally the worst polluters, are retired. As Figure 2-4 shows, however, this will take many years; and great pressure has been brought on the Congress to accelerate these reductions through stringent controls on older power plants -- those not covered by the existing CAA.

The National Acid Precipitation Assessment Program (NAPAP), in its 1988 Interim Assessment, reported that NO_x emissions increased in the United States over the period 1900 to 1950, held relatively steady

Figure 2-4

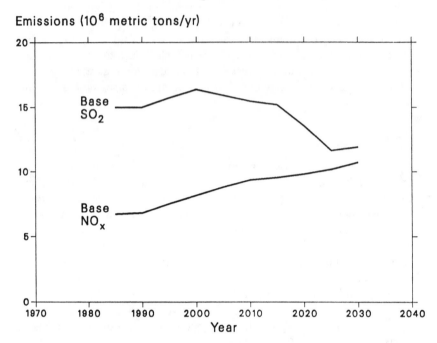

Emissions (10^6 metric tons/yr)

Source: NAPAP Interim Assessment, Volume II, 1987

through the late 1970s, and appear to have dropped slightly since 1980. Table 2-1 shows estimated man-made emissions of NO_x in the United States, by source for 1980 and 1985.[29] Most areas of the U.S. are in compliance with existing EPA standards for NO_x emissions. However, a study by Argonne National Laboratory, under contract to DOE, forecasts increases in NO_x emissions by approximately 30 percent between 1980 and 2010.

Transportation activity is the largest source of NO_x emissions, contributing 7.5 million tons per year in 1984. NO_x emissions are projected to remain relatively constant through 1995, and then grow to 9.3 million tons per year by 2010. Existing vehicular emissions controls are thought to be responsible for decreases that occurred between 1978 and 1983, and for the projected steady rate of emissions through 1995. By that year, however, the continuous rise in transportation demand (as well as

Table 2-1:   Estimates of man-made NO_x emissions from various source categories in the U.S. (million metric tons/yr. of NO_2)

	1980	% of Total	1985	% of Total
Highway Vehicles	7.2	37%	7.1	37%
Power Plants	5.8	30%	6.3	33%
Industrial				
Fuel Combustion	3.0	16%	2.9	15%
Other[a]	3.3	17%	3.0	16%
Total[b]	19.3		19.3	

[a]Includes residential and commercial fuel use, industrial processes, aircraft, railroads, vessels, and off-highway vehicles.
[b]Does not include natural sources such as lightning, soil emissions, stratospheric injection, and oceans. These are estimated at roughly 3 million metric tons/yr. of NO_2 equivalent emissions in North America.

Source: NAPAP Interim Assessment, (Volume II), page 1-49.

projected increases from industrial activity) is projected to overtake improvements that have resulted from existing control requirements under the Clean Air Act.

Emissions of NO_x from coal-fired power plants also are projected to rise, from 5.8 million tons per year in 1980 to 9.8 million tons per year in 2010, a 71 percent increase. This projection is based on increased coal use by utilities, and on new power plant construction projected to occur over this time frame. In fact, NO_x emissions from oil and gas-fired power plants until recently were expected to decline. However, this study did not consider the recent trend among utilities toward increments of gas-fired capacity that have occurred over the past few years and are expected to continue as long as oil and gas prices remain low.

Emissions of SO_2 and NO_x have begun to have an impact in the U.S. Increased acidity is certain. Monitoring data indicate that nearly the entire U.S. east of the Mississippi (excluding the deep South) receives precipitation that is as much as 10 times more acidic than it would be if falling through unpolluted air.[30] In 1984 the Office of Technology Assessment reported that approximately 3,000 lakes and over 23,000 miles of streams were then "extremely vulnerable to further acidic deposition or have already become acidic."[31] Acidic deposition has taken its toll, primarily on fish populations such as those in the Adirondacks, Ontario and Scandinavia, but the long term effects of increased acidity go beyond population. Under current rates of deposition a long-term pH of 4.9 or less can be expected for low alkalinity lakes and streams in the Northwest and Canada, a pH level that can eliminate most fish species, nearly all mollusks and many forms of aquatic life.

Upon reaching the earth, acid rain usually affects life directly, simply by causing intolerable levels of acidity in lakes and streams or by mobilizing toxic metals already existing in the water supply. Trace metals, such as mercury, cadmium, and aluminum, are all released into the water by the presence of sulfuric or nitric acid. The implications for indirect human health effects are clear. As toxic metals are released into the water, not only is the drinking water contaminated, but food sources then may contain the toxic chemicals as well.

There is strong circumstantial evidence that acid deposition is involved in a widespread decline of tree growth in many parts of the world, and an increase in tree mortality. Either by directly removing the nutrients from tree canopies or by damaging the soil, acid rain has the potential to do vast damage to forests. Trees covering at least 10 million hectares in Europe, an area the size of East Germany or the state of Ohio, now show signs of injury linked to air pollutants or acid rain. A quarter to half of the forest area is damaged in the countries of Austria, Czechoslovakia, Luxembourg and the Netherlands. In autumn 1983 the West German ministry of Food Agriculture and Forestry found that 34% of the nation's trees were yellowing, losing needles or leaves or showing other signs of damage. By 1986, the German government reported damage to 54% of their forests. Preliminary findings pointed to air pollution and acid rain as contributing if not causal factors.[32] Researchers in that country project from trends that forest damages in their country over the coming decades will average $2.4 billion/year including losses to forest industries and water quality and recreational values.[33]

Global Warming

Fossil fuel combustion plays a large role in a third area of environmental concern that has begun to receive widespread attention. It has been hypothesized that the buildup of trace gases in the atmosphere as a result of human activity will eventually lead to a significant increase in global temperatures. Carbon dioxide and several other gases (primarily methane, chloroflourocarbons, tropospheric ozone, and nitrous oxide) let energy from the sun pass through the atmosphere but absorb longer wavelength radiation reflected from the earth's surface. The process is similar to that which occurs in a greenhouse, and is commonly called the "greenhouse effect."

The release of carbon in the form of CO_2 is an essentially inevitable result of fossil fuel burning with current technologies. Although CO_2 is a common gas--it is released every time we breathe--its atmospheric concentration has increased substantially since the Industrial Revolution in the 1850s. The more recent destruction of tropical rain forests has exacerbated the situation because these forests absorb large quantities of CO_2 as part of the process of photosynthesis. Plants and the Earth's oceans absorb CO_2 emissions, but we appear to be outpacing the natural carbon cycle by releasing unprecedented levels of carbon into the atmosphere. It has been estimated that only about half of the carbon presently being emitted is absorbed. The greater the concentration of greenhouse gases, the more heat that is trapped.

By analyzing air bubbles trapped in glacial ice, it has been established that atmospheric CO_2 levels, over the centuries, have averaged about 280 parts per million. This was the approximate level at the beginning of the industrial revolution in the mid-19th century. Between the years 1959-85 however, the amount of CO_2 in our atmosphere has climbed from 315 to 345 parts per million, and continues to climb by roughly one-half of one percent a year.[34] Recent calculations show the combustion of fossil fuels in 1985 increased the atmospheric carbon dioxide content by 11.4 tons, higher than in any previous year.[35] It has been estimated that atmospheric concentration of CO_2 will reach double its pre-industrial levels between the middle and end of the next century.

It is important to note that the warming component from greenhouse gases *other than* CO_2 has become quite large in recent years. In the 1980-85 period, the annual contribution to the greenhouse effect from these other gases has almost equalled that of CO_2. Among the six primary greenhouse gases noted above, the two CFCs have by far the greatest warming effect per molecule. Control of these other gases, particularly CFCs, will have a major impact on whether the temperature effects of a CO_2 doubling in the atmosphere will occur earlier or much later in the next century. Concentrations of other greenhouse gases in 1980 were equivalent to another 40-50 ppmv of CO_2. It is projected that by 2030, the equivalent concentration of these gases will be about 140 ppm. of CO_2, assuming no changes in current trends.[36]

Based on current understanding of interactions between the atmosphere and climate, the prevailing consensus within the scientific community is that when the amount of trace gas concentrations in the atmosphere reaches a point that represents an "effective doubling" of CO_2 over its pre-industrial level, the earth will be comitted to an eventual warming of between 1.5 and 4.5 degrees Centigrade. Most believe this point will be reached by about 2030 if present rates of emissions continue.[37] It is important to note, however, that the actual warming effects lag this point by anywhere from 11 to 50 years.

Many scientists see the rise in the earth's temperature as the most serious concern among current environmental problems. Estimates by scientists in the U.K. have concluded that global warming from the pre-industrial period until now has been from 0.3 to 1.1^0 C (0.5 to 2.0^0 F). This may not, at first seem significant, but to scientists who study climate it has been the cause of considerable attention and research. As a point of comparison, the earth's mean temperature is only 1^0C higher than it was during the "Little Ice Age" of the 13th to the 17th centuries. This extended period of low temperatures resulted in the abandonment of about half the farms in Norway, an end to the cultivation of cereals in Iceland, and some farmland in Scotland being permanently covered with snow. Whether or not equally dramatic effects will result from a one degree *warming* is uncertain.

While almost all authorities agree that the increase in greenhouse gas concentrations will lead to some resultant warming, there is disagreement over both the levels of warming and the time frame. One of the more disconcerting forecasts has come from James Hansen, director of NASA's Goddard Institute for Space Studies. He projects that, as soon as 20 years from now, global temperatures should be nearly 2°F higher, which would be "the warmest the Earth has been in the last 100,000 years."[38] Less pessimistic estimates are for later in the 21st century, and the Environmental Protection Agency has estimated an increase in average temperature of about 2-8 degrees Fahrenheit.[39]

The effects of such a rise in temperature go beyond just a warmer climate. As the atmosphere warms the ocean, thermal expansion occurs: the warmer water expands and takes up more volume, causing a rise in the sea level. Simultaneously, the warmer atmosphere and warmer water cause melting of the polar ice cap. Because water is less dense than ice, this causes the sea level to rise further. A rise in temperature that falls into the upper end of the predicted range could cause the sea to rise three feet or more, putting many coastal areas under water.[40] An October 1986 report by the United Nations Environmental Program and the U.S. EPA predicted that the oceans could rise 2 to 6 feet by the year 2100 as a result of the greenhouse effect.[41] This change could cause wide destruction along the Atlantic Gulf Coast, and drinking water supplies could be contaminated by salt water.

Beyond the coastal erosion and possible water contamination, the rise in temperature and subsequent sea level rise could have a dramatic effect on our climate. Heat waves would occur with greater frequency, and according to Jim Hansen, with the doubling of carbon dioxide levels, Washington would get 87 days a year with temperatures above 90 degrees, compared to 36 days now.[42]

Other climate models suggest rainfall patterns could shift, drying out the Midwest by 10-20%, reducing corn, wheat and other staple crops, and many believe the increasingly severe droughts that have afflicted the midwest in recent years provide a foretaste of what conditions may be like under the worst greenhouse scenarios.[43] One model, by an atmospheric scientist at Princeton University, forecasts drought patterns comparable to those of the Dust Bowl Era of the 1930s.[44] While highly uncertain, some estimate the annual cost of climate change could approach 3% of the world's gross economic output, perhaps cancelling the benefits of economic growth.[45]

Although there remains a significant degree of uncertainty associated with current forecasts of a substantial global warming, the potentially serious effects of increasing concentrations of "greenhouse gas" emissions on human welfare suggest the need to pay attention to the issue while we may. Because a substantial lag time exists before the effects of these emissions may be felt, and because of the long atmospheric lifetimes of critical pollutants, to wait for definitive proof on the global warming hypothesis would be an unnecessary risk. The need for further research now, as well as the need to develop mitigation strategies with respect to greenhouse gas emissions, could not be more clear. Most importantly, the role of fossil fuels combustion in the greenhouse effect makes clear the need for alternative fuels as a means of reducing a potentially serious threat to our environment.

Chapter Two: Summary

Coal Use

Coal is the most abundant fossil fuel. Proven reserves in the U.S. alone approach 500 billion tons -- an amount that would supply domestic needs in the U.S. for several hundred years at current consumption levels -- and recoverable reserves world-wide hold about three times as much energy as oil and natural gas reserves combined.

For much of U.S. history, coal was our primary fuel source. By the beginning of the early 20th century, however, coal began to lose dominance as a result of the expansion of gas and oil pipelines across the U.S. Although coal is low-cost and abundant, it has its disadvantages. Its bulk makes it difficult to transport and store; transportation costs frequently account for one-third of the price of delivered coal. In addition, the heat content per unit of volume of coal is significantly lower than either oil or natural gas. One ton of coal is required to produce the same amount of heat as only four barrels of oil.

Two other problem areas, however, present the most difficult hurdles for increased coal use. First, changes in the U.S. energy economy have led both utility executives and financial markets to view investments in large, capital intensive power plant construction projects as extraordinarily risky ventures. Second, coal burning produces serious levels of several pollutants implicated in air quality problems, acid rain, and global warming. Although there are similar environmental problems associated with the combustion of oil and natural gas, the effects of coal-burning are both more severe and more expensive to control.

Environmental Impacts of Fossil Fuel Use

Dramatic increases in the use of fossil fuels since the Industrial Revolution of the 1850s have resulted in serious environmental consequences. Fossil fuel combustion results in emissions of various gases and other pollutants that raise three key areas of environmental concern: effects on ambient air quality, acid precipitation, and global warming (the "greenhouse" effect).

The most severe air quality problem, in terms of the number of Americans affected, is tropospheric (ground-level) ozone. Ozone acts as a respiratory irritant, aggravating heart and lung diseases. In 1986, 75 million people lived in areas where federal air quality standards for ozone levels were exceeded. Most other pollutants that define ambient air quality have shown improvement due to Clean Air Act restrictions.

Acid rain results primarily from emissions of SO_2 and NO_x, which react in the atmosphere to form sulfur and nitric acid and are carried to earth in rain water. Acid rain affects ecosystems directly by increasing the acidity of watersheds, and catalyzes reactions of metallic substances in lake and river beds, resulting in potentially serious human health effects. There is also strong circumstantial evidence that acid deposition is involved in a widespread decline of tree growth in many parts of the world. Emissions of acid rain precursors come chiefly from coal-fired power plants and vehicular exhaust.

Global warming is hypothesized to result from the buildup in the atmosphere of several trace gases released by human activity. As "greenhouse" gases accumulate in the atmosphere, they act as a one-way filter, allowing sunlight to reach the earth's surface, but trapping heat reflected back toward space. The most important energy-related greenhouse gas is carbon dioxide. Based on current understanding of interactions between the atmosphere and climatic processes, most scientists who focus on this area believe the earth may warm by 1.5 to 4.5 degrees Centigrade by the middle of the next century. There is disagreement over the level of warming earth may experience, and in what time frame. However, the effects of warming ultimately could be quite serious, including a rise in sea-levels (threatening many coastal cities), altered climatic zones, and an increase in the frequency of extreme meteorological events such as heat waves, hurricanes, and droughts.

The role of fossil fuels in each of these environmental threats makes clear the need for alternative fuels as a means of averting potentially serious consequences for human health and well-being. Moreover, because the magnitude of these threats generally is expected to increase with time, we would do well to find energy alternatives sooner, rather than later.

Chapter Two: Vital Statistics

Coal Use

Demonstrated coal reserves in the United States: approximately 500 billion tons.

Coal accounts for 90% of total U.S. energy reserves, but provides only 23% of energy used.

In 1987, electric utilities accounted for almost 85% of coal consumed in the U.S.

Electricity load growth from 1973 to 1982: about 2.5% per year, down from an average of roughly 7% per year over the last few decades.

Emissions

U.S. sulfur emissions in 1980: 23.6 million metric tons (MMT). Utility power generation accounted for 15.9 mmt, of which 14.7 mmt came from coal-fired utilities.

Utility sulfur emissions in 1986: 14.4 mmt from U.S. plants.

Five states -- Missouri, Illinois, Indiana, Ohio, and Pennsylvania -- account for more than 40% of utility sulfur emissions.

Cost per ton for selected levels of sulfur
emissions reductions in 1985 dollars:

Reduction	Fuel Switching?	
	Yes	No
8 mmt/year	$ 270	$ 306
10 mmt/year	360	528
12 mmt/year	779	-

U.S. NO_x emissions in 1980:   19.4 mmt, of which 5.8 mmt came from utilities. Transportation sources accounted for a larger share, at 7.2 mmt.

U.S. NO_x emissions in 1986:   19.1 mmt., with utility emissions increasing to 6.3 mmt.

Global man-made CO_2 emissions in 1860: 93 million tons.

Global man-made CO_2 emissions in 1984: 5 billion tons. Transportation sources account for 700 million tons, or about one-third of man-made emissions in the U.S.

The average American car pumps its own weight in carbon into the atmosphere each year.

Environmental Concerns

Number of Americans living in areas where ozone levels exceed federal ambient air standards: 75 million.

Precipitation received in the U.S. is 10 times as acidic as it would be if falling through unpolluted.

Projected global warming from an effective doubling of atmospheric CO_2 concentration: 1.5 to 4.5°C.

Chapter Two: End Notes

[1]*Energy Security, op. cit.,* p. 165

[2]*Ibid.,* p. 43.

[3]*Coal - Bridge to the Future,* Report of the World Coal Study, 1980, p. 161.

[4]*Energy Security, op. cit.,* p. 166.

[5]*Economist,* August 8, 1987, p. 89.

[6]*Monthly Energy Review, op. cit.,* p. 7.

[7]*Energy Security, op. cit.,* pp. A-15 to A-21.

[8]*Ibid.,* p. 171.

[9]R. E. Disbrow in remarks to the 1988 Nuclear Energy Forum, November 2, 1988, Washington, D.C.

[10]National Acid Precipitation Assessment Program (NAPAP) *Interim Assessment, Volume II, Emissions and Control,* Washington, USGPO, 1987, p. 1-25.

[11]NAPAP 1987 Annual Report to the President and Congress, Washington, USGPO, 1988, p. 11.

[12]Yergin and Stobaugh, p. 86.

[13]Congressional Budget Office, *Curbing Acid Rain: Cost, Budget, and Coal-Market Effects,* Washington: USGPO, June 1986, p. xvii.

[14]*Ibid.,* pp. xix-xxiv.

[15]*Ibid.,* pp. xxiv-xxvi.

[16]Postel, Sandra, "Altering the Earth's Chemistry: Assessing the Risks," (Washington, D.C.: Worldwatch Institute) July 1986, p. 8.

[17]Flavin, Christopher and Alan B. Durning, "Building on Success: The Age of Energy Efficiency," (Washington, D.C.: Worldwatch Institute) March 1988, p. 23.

[18]Electronic Power Research Institute (EPRI) Study, findings presented at Conference on Global Atmospheric Change Linkages, Washington, D.C., November, 1988.

[19]EPA, 1988: NAAQS Trends Report.

[20]*Ibid.*

[21]U.S. Environmental Protection Agency, "Air Quality Benefits of Alternative Fuels" (Washington:GPO), June 1987, Table II.

[22]Postel, *op. cit.,* p. 34.

[23]U.S. Environmental Protection Agency, 1988: National Air Quality and Trends Report, 1986, page 3-16.

[24]Andy Pasztor, "U.S. Wants to Combat Gasoline Fumes From Cars Refueling At Service Stations," *Wall Street Journal* (July 23, 1987).

[25]Robert E. Taylor, "EPA Deals Blow to Auto, Truck Makers in Ruling on Handling Gasoline Fumes," *Wall Street Journal* (July 23, 1987) p. 16.
[26]*NAPAP Interim Assessment: The Causes and Effects of Acidic Deposition*, Volume II, September 1987, p. 1-25.
[27]U.S. Office of Technology Assessment, "Acid Rain and Transported Air Pollutants: Implications for Public Policy," June 1984, p. 9.
[28]*NAPAP, op. cit.*, p. 3-11.
[29]*Ibid.*, pp. 1-45.
[30]U.S. Office of Technology Assessment, *op. cit.*, p.9.
[31]*Ibid.*
[32]Postel, *op. cit.*, p. 6.
[33]*Ibid.*, p. 21.
[34]The World Commission on Economic Development, *Energy 2000: A Global Strategy for Sustainable Development:* Report to the World Commission on Environment and Development (WCOED, NJ) 1987, p. 24.
[35]Jeff Nesmith, "Increasing CO_2 Levels Worry Scientists," *Palm Beach Post* (September 7, 1986)
[36]*Ibid.*
[37]National Academy of Sciences Report, 1987.
[38]James E. Hansen, testimony before U.S. Senate, August 1988.
[39]"Study: Pollution Warming Climate, Destroying Farms," *Palm Beach Post* (July 20, 1986), p. 8A., and Robert E. Taylor, "Policy Makers, Spurred by Ozone Treaty, Consider Tackling Greenhouse Effect," *Wall Street Journal,* (September 17, 1987) p. 38.
[40]Taylor, Robert E., "Policy Makers, Spurred by Ozone Treaty, Consider Tackling Greenhouse Effect," *Wall Street Journal,* September 17, 1987, p. 38.
[41]Robert Thomas, "Future Sea Level Rise and its Early Detection by Satellite Remote Sensing," *Effects of Changes in Stratospheric Ozone and Global Climate, Volume 4: Sea Level Rise,* UNEP and U.S. EPA, October 1986, p. 28.
[42]*Ibid.*
[43]Hansen, Testimony before U.S. Senate, August 1988. Also, Daniel Dudek, Environmental Defense Fund, Testimony before U.S. Senate, August 1988.
[44]Doig, Stephen K., "Future Shock: Burger Toxes, Cars, Even the Fridge are Warping our Climate," *Miami Herald,* October 11, 1987, p. 16A.
[45]Postel, Sandra, July 1986, p. 6.

CHAPTER THREE

ALTERNATIVES TO FOSSIL FUELS

. . . the political fallout from Chernobyl will be its lasting legacy. Since the accident the pro-nuclear consensus has collapsed in country after country, and the future of nuclear power, already hanging by a thread in some nations, is now in greater jeopardy than ever.

-Christopher Flavin

NUCLEAR POWER

History

The first nuclear reactor was constructed in the U.S. by a group of scientists, headed by Enrico Fermi, and located at the University of Chicago. This "atomic pile" was a research reactor, constructed in a converted squash court under the football stadium; and it was, literally, a pile of graphite blocks with uranium fuel rods inserted among them--a very primitive experiment to test the feasibility of controlling a nuclear chain reaction.[1] Story has it that the neutron-absorbing control rods to shut-down the nuclear reaction were suspended in such a manner that they would fall into place if a single rope were cut by a man who was positioned with an axe. This insight into limits on early understanding of nuclear safety also provides the origin of the most common term for an emergency shut-down of a nuclear reactor, SCRAM (System Control Rod Axe Man).

The Chicago experiments were focused on the military uses of controlled nuclear energy, which culminated in the nuclear submarine programs of the Navy. The effectiveness of this new power source was demonstrated in the development programs that led to the launching of the USS Nautilus. The beginning of the commercial nuclear power industry in the U.S., however, is generally marked by President Eisenhower's "Atoms for Peace" speech, which he gave to the United Nations on December 8, 1953. This was followed in 1954 by amendment of the Atomic Energy Act to permit and encourage private industry participation in the development and utilization of atomic energy for peaceful purposes, particularly electric power generation. Congress followed this legislation with its five-year, five reactor program, which was authorization to build five industrial prototype reactors, the largest of which was the 60-megawatt reactor at Shippingport, Pennsylvania, completed in 1957.[2]

Shippingport was the first commercial reactor to generate electricity for the power grid. It used uranium fuel, formed into rods and placed in a steel vessel surrounded by water at very high pressure. The water was used to slow the nuclear fission process and carry the heat from that reaction through a closed-loop system of pipes, into a steam generator, which was connected to a steam turbine (Figure 3-1). The nuclear reactor at Shippingport was the first pressurized water reactor (PWR) constructed in the U.S., and this has become the most common technology in use among U.S. utilities. The second major technology used by U.S. utilities is the boiling water reactor (BWR), which also uses water to moderate the nuclear reaction but operates at a much lower pressure, with the steam generated in the reactor core used directly to drive the turbine-generator. Both of these water-moderated forms of nuclear power generation are generically called "light water" reactors, to separate them from a form of the technology that uses deuterium oxide, or "heavy water," to moderate the fission reaction.

Several PWRs and BWRs were ordered in the late '50s and early '60s, but all had some form of subsidy from the U.S. Atomic Energy Commission (AEC) until December 1963, when Jersey Central Power and Light ordered a 515 MWe plant from General Electric, to be built at Oyster Creek under a "turnkey" contract. This meant that GE undertook to deliver a completed nuclear power plant, ready to operate, for a fixed cost, to be adjusted only by inflation. Three other major manufacturers began to offer completed generating stations at firm prices. These were Westinghouse, Combustion Engineering, and Babcock and Wilcox.[3] These four companies formed the nucleus for virtually all commercial nuclear power plant construction in the U.S.; and the late '60s and early '70s became a period of intense competition for the reactor market among these four manufacturers. Nuclear power was considered to be the future of the electric utilities industry--power that was "too cheap to meter"--and there were projections that there would be 300,000 to 500,000 MWe of nuclear capacity by 1990.[4] The number of nuclear plant orders grew rapidly, peaking in 1973 with 41 orders for new nuclear units in that year alone.

Figure 3-1

PRESSURIZED WATER REACTOR SYSTEM

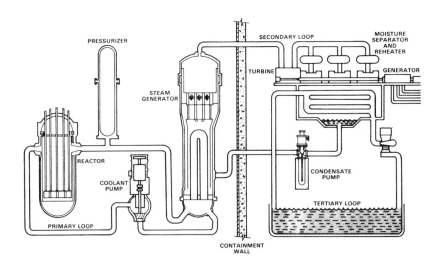

Schematic of a PWR Plant Design

Source: U.S. Nuclear Regulatory Commission

The heyday for nuclear power extended through 1973; but in the mid-1970s, the picture changed rapidly. In 1974, 28 new plants were ordered; and another 13 were ordered in 1975-1978. All 41 of those plants were cancelled, and 32 of the 41 that were ordered in 1973 were cancelled. Of the 249 commercial nuclear plants that have been ordered in the U.S. since 1953, 119 have been cancelled.[5] Before embarking on a discussion of the reasons for such a turn-around, however, it is important to look at the role nuclear energy plays in the electricity sector of the U.S. economy.

Electricity Generation

At the end of 1988, the U.S. had 109 reactors licensed to operate, with a total power generating capacity of 97.2 MWe. Another 14 plants, totalling 16.6 MWe, were under various stages of construction, and two others remained on order. Of the licensed plants, 101 were in operation; and these plants provided about 20% of total U.S. power generation in 1988. Nuclear is second only to coal in total annual electric power generation, providing twice the amount produced by either gas or hydroelectric and about four times the amount produced by oil.[6] This level is expected to continue to grow as plants still under construction come on line and those plants with operating licenses that have been shutdown come back into service.

Several states in the U.S. have become quite dependent on nuclear power. One-third of New England's power comes from nuclear generation. Fifty percent of the electricity in the states of Connecticut, New Jersey, South Carolina and Vermont is nuclear power; and another 12 states are dependent on nuclear for more than one-fourth of their electricity. There currently are nuclear plants operating in 33 states.[7]

It also is informative to place the level of U.S. nuclear power generation in the context of worldwide use of nuclear energy. More than 400 reactors are in operation worldwide in 26 countries, and nuclear power provides about 16 percent of the world's electricity generation. More than 130 additional reactors are under construction. The U.S. has the largest number of reactors of any nation, followed by the Soviet Union with 57, France with 49, Great Britain with 38 and Japan with 37. The U.S. will maintain its lead in the *number* of reactors for the foreseeable future, although many nations will continue to have a much larger share of their electricity generated from nuclear.[8]

Economics of Nuclear Power

It is generally believed that the demise of the nuclear power revolution in the U.S. is a result of safety problems. There is a perception that the combination of the accident at Three Mile Island and the recent Chernobyl accident have heightened public fears of the possibility of a major nuclear disaster and have simply created an aura of danger that makes nuclear power unacceptable to the public. There is truth to the sharp drop in public acceptance of nuclear power. Polls have shown a steady decline in public attitudes toward the construction of new nuclear plants, and the acceptance of nuclear power hit an all-time low following the 1986 accident at Chernobyl (Figure 3-2). Attitudes reflected in polls do not always translate into actions that would be consistent with them, however, especially when public attitudes clash with economic self-interest. This has proved to be the case when the public has been presented with numerous opportunities to vote to close nuclear power plants over the past few years. Through 1988, there have been 15 referenda on state ballots to close nuclear plants; and in all cases, the public has decided that they should continue in operation, primarily because costs of alternative power sources were higher.

The key point is that economics of nuclear power generation plays a critical role in public attitudes and acceptance of nuclear power. The 1987 DOE study of energy security issues identified three major factors that influence public attitudes toward nuclear power: the potential for a severe accident; the seeming inability to find a way to dispose of nuclear waste in a safe and permanent manner; and increasingly poor economics of the nuclear power alternative.[9] Experience with public referenda on nuclear power may suggest that when push comes to shove, the cost impacts of the nuclear option will hold sway, particularly when consumers can visualize the direct impacts on their utility bills. This can cut two ways, as the controversy over the Shoreham nuclear power plant has shown; and utility consumers have become much more sophisticated in their capabilities to influence such decisions.

The earlier discussion of commercial nuclear power development shows that nuclear construction was experiencing some economic difficulties long before the accident at Three Mile Island. New plant orders declined sharply after 1974, while public acceptance of nuclear power was still quite high. This was not a problem with nuclear power as an electrical generating technology, but instead marked a watershed in the traditional way that utilities had identified and addressed their needs for power. The underlying reason for the initial change was the post-oil embargo jump in electricity prices and the accompanying sharp decline in electricity demand. Delays in plant construction and a period of high inflation caused prices to rise even higher and kept demand growth down. Retail prices for residential users of electricity doubled between 1973 and 1980; and by 1984, prices had tripled over 1973 levels. For industrial users, prices had doubled by 1977, tripled by 1980 and quadrupled by 1984.[10] There is little doubt that for electricity users of all types, this was a new and difficult economic environment, in which they were forced to adjust downward their demand for electricity.

The Chapter 2, Coal section discussion of the economic environment for electric utilities points out that electricity load growth

Figure 3-2

Shift in Public Attitude Toward Nuclear Energy

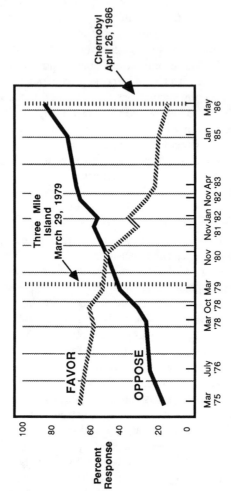

Answers to the question: "In general, do you favor or oppose the building of more nuclear power plants in the United States?"

SOURCES: 1975-1980 ABC News-Louis Harris and Associates, Inc.,
1981-1982 NBC News, 1983-1986 ABC News-The Washington Post

Published in *Energy Security: A Report to the President of the United States*, U.S. DOE, March 1987.

had been almost a constant 7% for several decades, a pattern that was sharply reversed in the 1973-1982 period, when it averaged only 2.5%. A quick calculation shows that a growth rate of 2.5% results over 10 years in a total growth of about 28%, while a 7% growth rate results in a 100% increase at the end of 10 years. Utility projections were forecasting the need to double electric generating capacity by the early 1980s; and due to long construction lead times for new plants, programs were well-underway when evidence of slower growth became apparent. Even then, one or two years of reduced growth was not enough to convince most utilities that their long-term forecasts were so far out-of-line with what the future would hold. The majority of plant cancellations did not take place until after 1977, and it was not until after 1980 that electricity demand forecasts began to drop below the 4% level to more closely match clearly observable trends.

As was noted in Chapter 2, the impacts of the past 15 years have been severe for new construction of both coal-fired and nuclear plants. Although coal-fired generation has recovered somewhat, there have been no new orders for nuclear plants since 1978; and *every* nuclear plant ordered after 1973 has been cancelled. It is almost a certainty that no commercial nuclear plants will be ordered, constructed and placed in operation by the end of this century if they are not now in the pipeline. A major question exists as to whether there will be any new nuclear plants *ordered* in the U.S. by the end of this century.

Coal and nuclear are both fuels for baseload plants that require large capital investments, and both are affected by the current economic biases (discussed in Chapter 2) against such utility investments. However, the problems facing nuclear power over the next several years are much more severe in many respects. The key differences that will influence the future of nuclear power in the U.S. fall generally into three categories: the turbulence and uncertainty of the regulatory environment (both Federal and state); problems with the technology; and deteriorating economics of nuclear power operations.

Nuclear Regulation

Regulatory changes have been the major cause behind delays in plant construction schedules. Nuclear power plant construction required only 4 to 6 years in the late 1960s, but those construction times have been extended to more than 15 years today. Although some regulatory changes and retrofit requirements were experienced before 1979, the volume increased dramatically following the accident at TMI. Orders for design changes were issued from several different offices within the staff of the Nuclear Regulatory Commission (NRC); and until very recently, there has been no requirement to assess such change orders for their total contribution to increased safety of the plant. The cost of these changes, both in capital investment and stretch-out of construction, has been huge. Interest and carrying charges resulting from schedule changes have been as much as 40% of the cost of several plants that recently have been completed. The worst case example is the Shoreham nuclear plant on Long Island, an 820 MWe plant that was estimated to cost about $500 million when started in 1968. The current cost estimate has exceeded $5 billion, and carrying costs alone have exceeded $3 billion.[11]

State regulation has grown to become more of a problem for nuclear power over the past decade than has Federal regulation. State public utility commissions (PUCs) make the final determination of a utility's need for power and what source that power should come from. After the plant is constructed, the PUC determines whether the plant can enter the rate base, how much of the construction costs can be charged to a utility's customers, and what overall rates will be for a particular service area. As noted above, retail electricity rates have increased sharply since the early 1970s; and this has resulted in consumers bringing pressure on State PUCs to control rate increases. The rapid growth in construction project costs resulted in many utilities seeking rate increases on an annual basis, and some more frequently.

A major problem that nuclear utilities have encountered in the past few years has been questioning from PUCs on whether plant construction expenditures have been "prudent" or not. PUC determinations that, for one reason or another, management decisions were wasteful or not in the best interest of their customers, have resulted in billions of dollars of cost recovery that were disallowed. This has taken two basic forms. One is not allowing a utility to place the costs of the plant into rate base when construction is complete, usually based on the logic that the plant was excess to the region's power needs and the utility should have recognized this fact and terminated construction before it progressed too far. A second prudence action is the disallowing of a part of the costs, based on poor management and evidence that the costs were imprudently incurred. Schedule extensions are a common basis for allegations of this type of imprudence. The Shoreham plant provides the worst case example of this, with $1,395,000,000 disallowed by the New York PUC as imprudently incurred expenditures.

The Federal and state regulatory environments for nuclear power are considered by the financial community (investment bankers, lenders, etc.) as likely to remain uncertain for an extended period of time. This means, from their point of view, that new projects cannot be completed within a specified time period with any degree of certainty. Therefore, it is virtually impossible for a utility to secure financing for new nuclear plant construction, even if it wanted to.

In a recent speech, former Secretary of Energy, James R. Schlesinger, stated what has become, among utilities, the prevailing view of state commission impacts on the nuclear power industry:[12]

> *The threat of bankruptcy is held out in front of any board of directors of any utility if it should have the boldness and the sense of the national interest to construct a nuclear plant. This is the reflection of a long history of regulation in this country that can be described as nothing short of a disgrace. Public Service Commissions of the states do not permit cost recovery for plants.*

Technology Problems

In addition to regulatory problems, the nuclear industry will have to address successfully a variety of technical problems with the current light water reactor technology. Current plants suffer from an extreme one-of-a-kind syndrome. Although reactor designs may be the same for a class of plants, almost every plant has many unique design elements that essentially "customize" it. This phenomenon has had an influence on everything from initial construction and plant maintenance to regulatory changes and utility emergency planning. A number of attempts have been made to develop standardized plant designs that can be licensed more easily by the NRC, but none of these has been completed. Standardized plant designs have been used in France, Japan and other nuclear nations; and the benefits this approach has brought to construction lead-times, maintenance, and licensing have been clearly documented. For example, PWRs with the same basic reactor design that is used in many U.S. plants take less than six years to bring on-line. This is in contrast to recently completed U.S. plants, which are averaging almost 15 years.

In the economic and safety environment that proposals for new plant construction will encounter, efforts to develop standard plant designs are likely to face two additional requirements that were not considered a few years ago. Size is now a prime concern, as utilities can no longer afford to order such large increments of power as the 1,000 MWe and larger plants that have become the standard for recent nuclear construction. In order to compete in this new market, plant designs may have to be sized in the 300 to 500 MWe range. Further, the newer technologies are expected to demonstrate the "inherently safe reactor" concept; that is, if something goes wrong, the reactor will shut-down automatically and safely. This combination of "modular" and safe technologies is on the drawing boards, but nothing is in the demonstration mode; and it is not likely that such designs can be completed and approved for licensing by the end of the century without substantial Federal support.

Nuclear Operations

The last key point in this discussion of nuclear power economics is the trend toward higher operating costs. Once a high initial investment was made in a nuclear plant, the expectation was that it would perform with a high level of reliability and at low operating costs. On average, nuclear plants have proved to be *less* reliable within the U.S. than other baseload plants; and the complexity of the technology has caused a range of unexpected problems for utilities, particularly those with only one or two nuclear units and lacking the trained resources to handle such problems in-house. A 1988 analysis of this issue, completed by DOE/EIA, shows that the trend toward higher-cost nuclear plant operations began in the mid-1970s, when real nonfuel operating costs for nuclear plants for the period from 1974 to 1984 grew at 12% per year, while comparable costs for coal plants grew at only 2% per year. Over the 10-year period, nuclear plant operating costs grew from $26 per kWe of capacity to $95 per kWe (1982 dollars). Industry performance has been utility and plant specific, with some breaking records for efficiency and low-cost operations while others have had extremely poor records of unexpected outages and low availability. The study found that problem plants, with higher than average civil fines, also had higher operating and maintenance (O&M) costs. This phenomenon, if it continues, will take away one of the alleged "benefits" of nuclear power.[13]

In summary, the dream of a form of electricity generation that was "too cheap to meter" lasted only a few years, and its passing left a very different future in its wake. Nuclear is now identified as both more costly to construct and more expensive to operate than coal, and the key economic factors that have placed it in this position are not likely to be reversed in the near future.

Health and Safety Concerns

The most often-cited concerns with nuclear power are potential safety and health effects. However, the likelihood of a severe accident with a large release of radioactive materials is quite small in the U.S. commercial nuclear power industry. In April 1985, the NRC reported to the Congress its estimate that there is a 45% chance of a nuclear core meltdown accident at a U.S. power plant within the next 20 years, based on an average estimated frequency of 3×10^{-4} (0.0003) accidents per reactor-year. This does not mean that there would be a radiation release associated with the accident. In the Three Mile Island accident, which is the only severe reactor accident that has occurred in the U.S., more than 50% of the core was melted; but the damage was contained, and the radiation released was negligible.

There is no NRC estimate of the probability of a large radiation release; however, the NRC does have a safety goal policy statement that was published in the *Federal Register* on August 4, 1986. This policy statement provides a guideline for the overall frequency of a large release of radioactive materials as 10^{-6}, or one in every 100,000 years of reactor operation. Whether a reactor is capable of meeting such an objective is determined through a detailed risk analysis of the plant systems, known as a probabilistic risk assessment (PRA). This approach tries to identify failure modes for each plant system and for the total plant. It cannot, however, identify every possibility; and in the two major accidents on record, TMI and Chernobyl, operator error played a major role. So far, nuclear power is more accident prone on a world basis than any of these projections of failure (Figure 3-3).[14]

Operational Safety

PRA is the best available approach to projecting the risks of a nuclear accident because there is such a small data base on which to use any other statistical measures to project such events. In terms of operational safety, however, the data do exist; and the American Medical Association (AMA) released a study in January 1989 comparing the health and occupational safety risks of nuclear power to other forms of electricity generation. This report found the risks related to nuclear power to be much lower than those associated with coal-fired power generation, stating:

> *If the coal is mined underground and transported by rail, the fuel cycle, from mining to combustion, is estimated to produce 279 illnesses and injuries, along with 18.1 deaths per gigawatt (10^9 watts)-year. In contrast, the nuclear fuel cycle, if the uranium is mined underground, is estimated to produce 17.3 cases of illness and injury and 1 death per gigawatt-year.*[15]

Figure 3-3

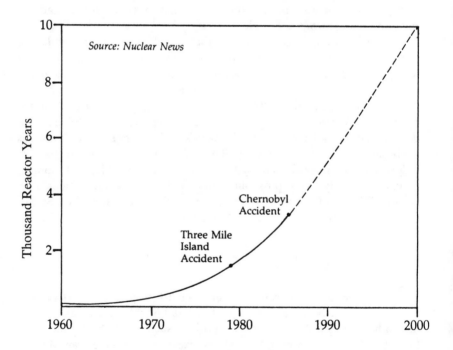

Cumulative years of Nuclear Reactor Operation Worldwide, 1960-85, with Projections to 2000.

Source: World Watch Institute, 1987.

The report went on to say that oil and gas-fired boilers, using current technology, are somewhat safer than those using coal or nuclear fuels.

Nuclear Waste

The other major concern most often expressed in terms of safety or risk is permanent storage of high-level nuclear waste. A decision was made in 1977 to postpone the reprocessing of spent nuclear fuel from LWRs. This meant that nuclear plants would operate on an incomplete, or "once through" fuel cycle (Figure 3-4). Spent fuel from commercial reactors would be held in storage until a permanent high-level nuclear waste repository was constructed and licensed. The Nuclear Waste Policy Act (NWPA) was passed in 1983, and the Federal government embarked on a 15-year program to site, construct and license such a repository for operation. The repository was originally scheduled to open in 1998, just about the time that several utilities would reach capacity for at-reactor storage of spent fuel. Four years into the program, the plan was revised; and the current year in which the repository is projected to open is 2003.

Whether such a repository can be constructed and licensed still depends on many unanswered technical, economic and political questions, given the requirement that in order to ensure safety, the effective life of the repository must be 10,000 years. In 1987, Congress passed amendments to the NWPA that specified the location to be Yucca Mountain, Nevada, unless the site investigation determined that site to be unsuitable. Since radiation release standards for the repository are projected 10,000 years into the future, this is as much a high-risk R&D project as it is a construction program. Further, it will require a first-of-a-kind NRC licensing process, complete with the full slate of public hearings and intervenors from a variety of interest groups.

The timely completion of the repository is regarded by both the Federal government and anti-nuclear forces as a necessary milestone in the continuation of nuclear power in the U.S. Therefore, every aspect of repository design, construction and licensing will be highly politicized; and there is no assurance that, without further Federal intervention into the process (possibly a Federal mandate), the repository will be completed on schedule--if at all.

If the repository cannot be completed on schedule, there will be a definite negative effect on any proposals for new nuclear power plant construction; and there is the possibility that some operating plants could be forced to close-down when they reach a certain level of spent fuel stored at the plant. Options to avoid this outcome include allowances for

Figure 3-4

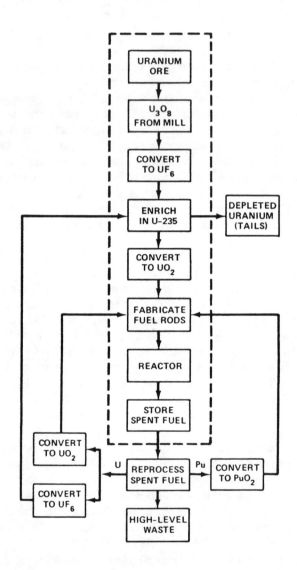

Uranium Fuel Cycle

Source: Energy Deskbook, U.S. DOE, 1982.

other forms of at-reactor waste storage, shipments to other plants that have more space, and development of interim storage sites. Currently, the law prohibits interim storage unless the final waste repository is under construction.

The Future of Nuclear Power in the United States

Forbes magazine, in a 1985 article, stated succinctly a prevailing view of nuclear power in the business and financial communities.

The failure of the U.S. nuclear program ranks as the largest managerial disaster in business history, a disaster on a monumental scale. It is a defeat for the U.S. consumer and for the competitiveness of U.S. industry, for the utilities that undertook the program and for the private enterprise system that made it possible. Without even recognizing the risks, the U.S. electric power industry undertook a commitment bigger than the space program or the Vietnam war.[16]

The likelihood of construction of any new nuclear plants by the end of the century, beyond those currently in the pipeline, is virtually nil. In fact, the likelihood of new nuclear plant orders in that timeframe is quite low. This is not to say that the scenario is impossible--no one who has watched energy markets over the past two decades should make such a statement. However, the revitalization of the U.S. nuclear industry in the next decade would require the confluence of several widely divergent factors. These would include: sharp increases in energy prices; high electricity demand growth; problems with other forms of electricity generation; solutions to Federal and state regulatory problems; solutions to problems with the technology; and probable government intervention to ensure the completion and operation of the first few plants.

Nuclear power will have a difficult time even being considered as a fuel source in an economic environment where oil, gas and coal remain low-cost options. The risks with nuclear are considered to be so great that they are likely to outweigh the fear of sharp oil and gas price increases that might result from an oil supply disruption. The logic to this argument is not that oil-fired generation will play an increasing role in the fuel mix--this is not the case. The price of natural gas, however, is tied to oil prices in the industrial and utility markets; and if oil prices increase, so will gas prices. A decade of decreasing oil and gas prices has engendered a level of complacency that has resulted in increasing orders for gas-fired electricity generation capacity, and it will take another price shock to force this market to add a premium for the risks associated with these fuels.

In addition to higher oil and gas prices, the favorable economic environment for nuclear power would require a high level of electricity demand growth; and the environmental costs of both current and future coal-fired generation, the principal baseload option, will have to be quite high. This would require both severe acid rain constraints and the potential for substantial global warming legislation, including probable restrictions on the generation of CO_2. At this point, the current level of science behind global warming projections does not provide support for draconian restrictions on most greenhouse gases.

Assuming the overall economic environment shifts to provide the above conditions, what is perhaps the most difficult hurdle must be addressed--the Federal and state regulatory institutions. Potentially the less difficult of the two is Federal (NRC) regulation. This will require the development of standardized plant designs that are pre-approved for licensing, one-step licensing, some insulation from the costs and delays of backfit requirements, and some relief from state and local vetoes of emergency planning arrangements. State regulation will be more difficult to address, given the 50 systems in the different states. It will be necessary to have an agreement with the PUC before construction starts that assures fair treatment of costs, both during and after construction, if the plant is built according to agreed specifications. The financial community, without such assurances mentioned above, simply will not provide the capital for construction. They cannot risk exposing investors to a replay of the events of the past 15 years.

All of the above assumes that the industry has managed to develop new forms of the technology that provide the safety and flexibility that the market will demand. This implies smaller increments of generation, "inherently safe reactor" technology, and development of the capability to construct a plant within a much shorter timeframe, perhaps 48 to 60 months.

The above review of the current status of the nuclear power industry, combined with an understanding of the elements that must come together for a revitalization of the industry, make it easy to understand why it is difficult to be at all optimistic about nuclear's future over the next decade or more. There is some potential for a resurgence based on the problems with fossil fuel combustion, and there are indications that some environmental groups are reassessing the conditions under which they could find the technology acceptable. There also are indications that, at least at the Federal level, the potential for a streamlined regulatory process is improving. However, the overall picture is not encouraging. It will take a major, coordinated effort by the nuclear power industry, and a substantial amount of serendipity, to navigate the industry through another reversal of its fortunes, to the point where there are new plant orders and construction underway. The more likely case, unfortunately, is that some part of the current backlog will be completed, and there will be a lengthy hiatus, perhaps for two decades or more.

RENEWABLES

In 1952 President Truman's Paley Commission warned of the need to conserve energy. Their report, *Resources for Freedom* foresaw fossil fuel shortages and strongly urged an aggressive solar research program.[17] Two decades later, with the advent of the first oil embargo, the interest in renewable energy sources was reborn. Solar panels appeared on many houses across the nation, including President Carter's White House, where solar collectors were installed on the roof for water heating.

In the late 1970s, total energy demand was high and growing rapidly. Energy efficiency in the U.S. was the lowest among the developed nations, with a virtual 1:1 relationship between GNP growth and energy growth. Oil prices were the highest in U.S. history; yet most forecasts of energy demand for the end of the century were projecting that the U.S. would be using more than 100 quads per year in the 1990s, a growth of more than 25%. In this high-cost energy environment, a wide range of programs to develop new energy sources were initiated. Federal programs for renewable energy R&D were funded at an all-time high, with budget proposals at the end of the Carter Administration approaching $700 million per year for renewable energy programs and another $700 million for conservation and energy efficiency programs.

The expected rapid growth in U.S. energy demand did not occur. In fact, a period of declining demand actually took place. Sharp increases in energy prices, combined with a strong push for energy conservation and efficiency, caused U.S. annual energy consumption to peak at 79 quads in 1979, with levels for the period 1981 to 1986 falling below the 1973 level of 74 quads. This change is more striking when one considers that it occurred while the U.S. population was expanding and the economy was increasing its production of goods and services. Between 1973 and 1986, the U.S. economy showed a rapid increase in energy efficiency. GNP grew 35% while energy consumption remained almost constant.[18] Yet as the energy crisis passed, so did the public's interest in both energy conservation and the development of alternative energy sources. Falling oil prices, combined with the phasing-out of Federal incentives for energy conservation and investments in development of alternative energy systems, resulted in a dramatic slowdown in renewable energy initiatives.

Resource Base

Hydroelectric power and combustion of biomass, primarily firewood, provide the largest contribution to the renewable energy resource base in the U.S. Precise accounting of the share of U.S. annual energy production from renewable energy sources is difficult because good statistics are kept only on hydropower and other renewable contributions to electric power sold to the grid. One estimate is that renewable sources provided 6.8-7.5 quads of U.S. energy in 1987, an amount equal to about 9 to 10% of our energy supply. Hydroelectric power accounted for about 3.1 quads, and biomass was estimated to account for another 2.6 to 3.5 quads.[19]

Renewable energy use falls generally into three categories: thermal applications (heating and cooling); fuels from biomass; and solar electric. Passive solar building construction and solar water heating are the two principal thermal applications that are in common use. The biomass category includes a variety of applications, such as grain alcohol for alcohol fuels, the cultivation of several herbaceous crops for energy production, gasified manure, etc. The solar electric category includes wind energy electricity generation, solar thermal electric (such as the "power tower"), photovoltaics (direct solar cell conversion to electricity), ocean thermal electric (using the differential in temperatures between ocean layers), geothermal (heat from sources deep in the earth's crust) production of electricity, and power from conventional hydroelectric.

With the exception of hydroelectric, most renewable use is found in decentralized applications and is difficult to measure. Passive solar use, for example, should be measured in terms of avoided energy costs for space heating and cooling. Industrial use of biomass is primarily restricted to wood waste in the pulp and paper industry; and residential use, although it has grown in recent years, is primarily as a supplementary fuel today in the U.S. Other technologies in the solar electric category are measured only if they contribute power to the grid. This misses many applications such as individual wind machines and solar installations on farms and residences. The current belief, however, is that beyond hydropower and woodburning, the total contribution to U.S. energy use is very small in relation to the more common commercial sources.

Hydroelectric Power

Hydroelectric power currently provides 6% of the energy consumed worldwide. In 1986 Venezuela completed the Guri dam, the largest hydroelectric plant ever, which with 10,000 MW capacity, can provide as much electricity as ten large nuclear plants.[20] Brazil and China are currently in the process of building plants even larger than this (15,000 MW).

The U.S. currently has the largest installed hydroelectric capacity in the world, providing 14-15% of our electricity use in a good water year.[21] Hydropower production is dependent on annual rainfall, and the best year in recent times was 1983, when over 3.5 quads of energy were generated in hydroelectric plants. This dropped to 2.6 quads in 1987 and was headed to a much lower level in 1988, with the summer drought. It is important to note, however, that hydropower production of electricity offsets the need for about three times as much energy from fossil fuels, given that the conversion efficiencies of fossil fuel power plants are much lower.

While only 59% of our hydropower potential had been tapped by 1980, it is unlikely that capacity will increase any time soon.[22] The major reason is environmental, as construction of dams for use as hydroelectric facilities can be extremely damaging to the environment. Much of the original hydropower construction was ancillary to other Federal water projects for irrigation, flood control and navigation; and the bulk of those projects were completed before World War II. Recent additions to generating capacity have primarily consisted of adding power houses to existing dams. The idea of constructing another Bonneville or Hoover dam project in the U.S. is not likely to receive serious consideration, given both the costs of such construction and the major environmental impact that such a project would have.

New technologies for small-scale, "low-head" hydroelectric generators (under 30 MWe) are being tested for use in rivers with good flow characteristics. According to DOE, there were 1,460 small-scale plants on-line at the end of 1985, totalling 6,800 MWe of capacity and generating about 0.4 quads of energy on an annual basis.[23] Although the estimates of additional capacity for such installations range from 10,000

MWe to much higher, the potential environmental problems with these facilities may well preclude the level of growth that has been hoped for. The New England River Basins Commission identified the ecological problems with this technology and estimated that only 3% of the 10,000 existing dams in New England could add small hydropower installations in an environmentally sound manner and remain economically viable.[24]

This limited future of hydroelectric growth in the U.S. has been recognized by the U.S. government. In 1987 the U.S. Bureau of Reclamation announced that its mandate to tap new water supplies had virtually expired and the agency would have to cut its work force in half in the following decade. The U.S. is relying increasingly on Canada for imports of power from their hydroelectric dams. About 15% of U.S. hydropower use in 1987 came from Canadian imports.[25]

Solar and Wind Power

Solar thermal energy and wind power generally have been considered to be the most promising of the newer renewable technologies.[26] Currently accounting for over 1,100 MWe of U.S. electricity production capacity, wind farms (huge fields of wind machines connected to the grid, such as the Altamont Pass, west of San Francisco) did not exist a decade earlier; yet this energy source saved the equivalent of about 1.5 million barrels of oil in 1985. The rapid success story of windpower however, is both a model for the potential of alternate energy programs and a case study for their failure.

The energy shortages of the 1970s inspired a wave of wind machine construction. From 1973 to 1983, over 10,000 wind machines were installed worldwide.[27] In the U.S., the state of California, with its many geographically appropriate wind passes, led the way. Wind installations grew from 144 machines in 1981, with a combined capacity of 7MW, to 16,769 turbines, with an installed capacity of 1,463 MW, by the end of 1987.[28] This amounted to just under one percent of the state's installed electric power generating capacity.[29]

Although windpower was cheaper than oil fired electricity during the energy shortages of the 1970s, banks were still reluctant to lend investment capital. For this reason, Congress passed the Public Utilities Regulatory Policies Act (PURPA), one of five parts of the National Energy Act of 1978, which enabled a wide range of newcomers, operating small cogeneration and renewable energy projects, to be certified by the Federal Energy Regulatory Commission as small power producers who could sell power to utilities at the "avoided cost" of the next increment of planned utility power generation. This meant that utilities were obliged to give priority to purchases of unlimited quantities of energy from PURPA qualifying facilities (QFs) at a rate equal to what the state utility commission determined to be the cost that the utility would have to pay for power if it had to construct a new plant. Investment in alternate energy projects was lucrative for high income taxpayers who were entitled to the usual federal tax credits for investment plus special federal and state credits for alternate energy investment. Together, these tax credits allowed consumers and businesses to deduct 40% of up to $10,000 in purchases of renewable energy equipment.[30]

In the first year the credits were offered, taxpayers spent $120.3 million on residential solar installations. By 1980 that number had risen to $399 million, and by 1981 it had grown to $678.6 million.[31] By late 1987, requests to construct more than 63,000 MWe of PURPA capacity had been granted approvals by FERC, equivalent to about two-thirds of our current nuclear capacity. The independent power industry is now doing $4 to $5 billion in business annually, and a 1987 study by the California Energy Commission found that many of these technologies operate more economically than conventional nuclear and fossil power plants.[32]

The response to PURPA initiatives has not been all positive, particularly from the utilities industry, which often finds itself forced to take expensive increments of power whether it needs it or not. As Richard Clarke, CEO of Pacific Gas & Electric, explained in an interview with the *Wall Street Journal* in 1987:

> *"PURPA requires us to buy power produced by so called independent power producers . . . even if we don't need it. It's a very expensive resource, we can't dispatch it and we can't count on it. So we have all these contracts for something like 9,000 MW. Our whole system has only a 15,000 MW capacity, so these contracts represent 60% of our energy capacity."*[33]

Even with incentives offered by PURPA, the decline in oil prices has once again offered low-cost, easily accessible supplies of energy. Since natural gas competes with oil in the industrial and utility sectors of the energy economy, gas has become the fuel of choice in the new utility construction market. PURPA has provisions for strong advantages to cogenerators, so gas is now used in a variety of cogeneration applications where electricity is the major output and process heat is secondary. The majority of new utility orders for capacity in the past few years have been for small to medium increments of gas-fired generation.

This, combined with the phasing-out of renewable energy tax credits and reform of the tax laws in general, has taken its toll on the renewable energy industry. The 15% federal tax incentives ended in 1986, and within less than two years the membership of the Wind Energy Association dropped from 100 to 40 businesses.[34] When the Solar Energy Industries met in 1986 only fifteen people showed up, representing an industry that once numbered 267 manufacturers, 6,000 distributors and 30,000 employees.[35] Energy investment credits for solar systems, restored at 12% in 1987 and 10% in 1988, expire in 1989. (Even the solar panels which heated the water in the White House for seven years following President Jimmy Carter's call for 20% of the U.S. energy needs to be met by renewable resources, were removed for cleaning in 1986 and never replaced.)[36]

The high expectations for rapid growth in the market for renewable energy supplies certainly were cooled with competition from unexpectedly low-cost oil and gas supplies. As with the synthetic replacements for fossil fuels, most renewable energy sources required high energy prices and some Federal subsidies to be competitive. When those incentives disappeared, only the ones that could compete head-to-head with cheap oil could survive in the market. Those that remain competitive are primarily forms of electricity generation that fit expensive (usually remote) needs for small increments of power. As continued R&D lowers the cost into a range that is competitive with utility power sources, the market will increase. This is the expectation for solar thermal and photovoltaics. Geothermal power has been very reliable for those regions where the resources are available. The current capacity of about 2,000 MWe could more than double by 1995.[37]

Two major conclusions can be reached from the discussion in this chapter that affect the thesis of this book. First, none of the non-fossil forms of electricity generation will provide an early alternative to continued, and quite likely increasing use of fossil fuels to meet electric power growth. The implications in this are: efforts to deal with the environmental aspects of fossil fuel combustion will be slowed as utilities hold on to older oil and coal-fired power plants; costs for electric power generation will increase to pay the costs of environmental controls that are mandated; or some combination of both.

In terms of liquid fuel substitutes for the transportation sector, that is, substitutes for petroleum, nothing approaching the economics of oil is on the horizon. Falling oil prices over the past decade have been a boon to the consumer, allowing prices for gasoline to return almost to pre-embargo levels and placing few economic restrictions on any transportation sources. These benefits to the consumer in the near-term have had an equally negative effect on programs to develop economic alternatives to petroleum products in the transportation sector. Although the U.S. currently uses much more oil than it produces every year in the transportation sector, the market provides no incentives to change this situation. It is a fallacy to believe that energy markets can provide anything more than an immediate indication of the levels of energy supplies and prices. Only a prolonged period of high prices, such as that which began to occur in the mid- to late 1970s, can affect the development of more expensive substitutes for fuel forms. In a period of low energy prices, only the most responsible government can provide the "public good" of looking to the future and attempting to assure a smooth transition to the inevitable future of high oil prices.

Alcohol Fuels

The two principal fuel alcohols are ethyl alcohol (ethanol) and methyl alcohol (methanol). Both can be produced from renewable resources and both can serve a variety of energy uses. Since the focus of this book is on methanol, that fuel will not be discussed in this section.

Ethanol

The most commonly recognized alcohol fuel is ethanol, known also as ethyl alcohol or grain alcohol. Produced from the fermenting sugars extracted from biomass, (most commonly corn and sugar cane), it has a molecular formula C_2H_5OH and an octane rating of 100. When mixed with gasoline (10%), it can boost the octane rating three points.[38] Ethanol is closer to gasoline in its physical, chemical and combustion characteristics than methanol. It has a higher energy content than methanol (about 2/3 that of gasoline vs 1/2), and generally enjoys the advantages of alcohol fuels while suffering less from their differences from gasoline.

Brazil has led the way in experimenting with ethanol as an automotive fuel. Its large feedstock of sugar cane, coupled with large foreign debt (which was aggravated by high imports of petroleum) precipitated the use of ethanol in this country. Used as a blending agent as early as 1933, ethanol became part of a national program to displace

Brazil's oil imports in 1976. By the early 1980s, all gasoline vehicles had to be capable of operating on a blend of 20% ethanol, 80% gasoline; and today 90% of new vehicles are powered by fuel-ethanol engines.[39]

Ethanol's greatest support in the U.S. has come from the farmers, who see it as an alternative market for their surplus crops. Production of ethanol for 1988 will utilize nearly 400 million bushels of corn in the United States, adding $1 billion to farm income while reducing federal farm program costs by $800 million.[40] Consequently ethanol enjoys the support of a large number of midwestern representatives. During the Arab oil embargo, gasohol, a blend of 90% gasoline and 10% ethanol, became a popular way to extend gasoline supplies; and the phaseout of lead as an octane enhancer has increased demand. In 1986, 785 million gallons of ethanol were blended into gasoline; and the industry has the capacity to blend more.[41] Estimates for 1988 project the production of 950 million gallons of ethanol in 80 production facilities located in 23 states.[42]

As a neat alcohol fuel, ethanol may not be considered superior to methanol; but its greater similarity to gasoline would probably win support from auto manufacturers and oil companies, were it not for practical considerations of price and supply. To replace 10% of the gasoline supply, some 5 billion bushels of corn or other feedstocks are required. Yet the total annual U.S. crop is only about 4.5 billion bushels.[43] The use of agricultural surplus for fuel production is an innovative response to several problems at the same time. Although the question of "food versus fuel" has been raised in the past, it is important to note that the food value of the feedstock actually can be enhanced in the ethanol production process. For example, when corn is used, only the starch (and often not all of that) is consumed in the conversion process. The protein value becomes more concentrated; and the resulting co-product, distiller's dried grain, is a very valuable food.

The major disadvantage of ethanol has been its cost. Although the cost of producing a gallon of ethanol in the most modern plants has dropped to approximately one-third (in constant dollars) the cost of only a decade ago, it is still not priced competitively in the marketplace. Various alcohol fuels initiatives appear to have locked ethanol into a pricing mechanism that forces it to compete with gasoline as a fuel supply extender rather than allowing a premium for its octane-raising characteristics. Ethanol is much better suited to the market as an octane enhancer since a 10 percent blend with gasoline adds about 3 octane points and requires no modifications to current engine technology.

Ethanol production costs can range between $1.15 and $1.60 per gallon, depending on the process used and the cost of the feedstock.[44] However, ethanol costs are falling while the cost of gasoline production fluctuates with the world price of oil and is expected to be much higher by the end of the century. The Department of Energy cost goal for ethanol by 1999 is 60 cents per gallon, based on technological improvements in plant design and manufacturing processes that have been developed and tested but not yet implemented on a commercial scale.

Ethanol prices reflect a complex web of Federal regulations and price controls. One reason for the current high cost of ethanol is the artificially supported prices of the primary feedstocks (cereal grains and molasses). These agricultural products are much more expensive than the competitive market would allow due to the broad range of farm price supports they receive. If these farm subsidies did not exist, it is likely that much of the cost differential between ethanol and gasoline could be eliminated, assuming stable oil prices and average conditions for world farm production. Instead, it has been found necessary to make ethanol prices more attractive through the use of an offsetting subsidy in the form of an exemption from the Highway Use Tax. Ethanol/gasoline blends (10% ethanol, 90% gasoline) receive a 6 cents per gallon exemption which is, effectively, a 60 cents per gallon subsidy for ethanol production.

Although on its face this exemption sounds like an excessive drain on the Highway Trust Fund, which the Highway Use Tax feeds, this is not the case for several reasons. First, this provision was enacted concurrently with an increase in the tax from 4 cents to 9 cents per gallon of gasoline; and the Highway Trust Fund has grown every year since the increase. This is growing much more rapidly than Congress has been willing to authorize expenditures. Second, only about 8 percent of current gasoline use is from blends eligible for the exemption; and this share would have to grow sharply before the impact would be felt on the Highway Trust Fund. Furthermore, the Congressional Research Service (CRS) has identified benefits from increased ethanol use in motor fuels that more than offset losses to the Treasury from the tax exemption.[45] This is primarily in farm program savings. Benefits that accrue from energy security cost reductions, reductions in the trade deficit and environmental advantages should add to these savings. However, the regulatory maze surrounding ethanol pricing makes this a complex calculation that is not immediately visible to the market.

It is conceivable that by the late 1990s, ethanol will be able to compete effectively in the motor fuels market as an octane enhancer, either in its alcohol form or as Ethyl Tertiary Butyl Ether (ETBE). It is not likely, however, that ethanol will be able to compete as a "neat" fuel with methanol. Methanol production costs already are substantially lower than ethanol; and methanol costs also will benefit from the same types of technological advances that should lower ethanol costs. Barring any unexpected limits on methanol availability, methanol should continue to have a clear cost advantage over ethanol.

Chapter Three: Summary

Nuclear

The beginning of the U.S. commercial nuclear power industry is marked by President Eisenhower's "Atoms for Peace" speech in 1953. The first commercial reactor to generate electricity for the power grid was completed at Shippingport, PA in 1957. Nuclear power was considered to be the future of the electric utilities industry -- power that was "too cheap to meter." The number of nuclear plant orders grew rapidly, peaking in 1973 with 41 orders. After 1973, however, the picture changed dramatically. All 41 plants ordered between 1974 and 1978 were cancelled, as were 32 of the 41 plants ordered in 1973.

At the end of 1988, the U.S. had 109 reactors licensed to operate, with a total power generating capacity of 97.2 MWe. Another 16 plants were under various stages of construction or on order. Operating nuclear plants supplied 20 percent of total U.S. electric power generation. In New England, one-third of the region's power comes from nuclear generation.

Public opinion polls have shown a steady decline in public attitudes toward the construction of new nuclear plants. However, economic forces have been key to the declining fortunes of the nuclear industry. Decreases in demand for electricity, together with increasing construction and operating costs for nuclear facilities, have led utilities to prefer other power sources. There have been no new orders for nuclear plants since 1978, and it is doubtful whether any new plants will be constructed -- or ordered -- before the turn of the century.

The economic picture for nuclear power contains elements beyond those which have made prospects bleak for new coal-fired plants. These include a turbulent regulatory environment, with resulting construction delays that pose cost overruns, and technological problems which have made plant operations unreliable. Finally, the health and safety problems associated with disposal of high-level waste remains thorny. To be safe, a waste repository would need to be secure for a period of 10,000 years. Whether such a repository can be constructed and licensed depends on many unanswered technical, economic, and political questions. The licensing process will no doubt be highly politicized, and federal intervention probably would be required for a site to be opened.

Leaving aside the significant problem of waste disposal, the revitalization of the U.S. nuclear industry in the next decade would require the confluence of several widely divergent factors. These would include: sharp increases in energy prices; high electricity demand growth;

solutions to federal and state regulatory problems; solutions to problems with existing plant technology; and probably government intervention to ensure the completion and operation of the first few plants.

Renewables

In the late 1970s total energy demand was high and growing rapidly. Energy efficiency in the U.S. was the lowest among the developed nations. Oil prices were at their highest level in U.S. history. In this environment, a wide range of programs to develop new energy sources were initiated. Federal programs were funded for renewable energy R&D at an all-time high, and tax incentives were put in place to encourage the use of renewable energy sources. In fact, however, expected rapid growth in energy demand did not materialize; and as the energy crisis brought about by the Arab oil embargo faded from public memory, so too did the public's interest in alternative energy sources. Falling oil prices, and phasing out of federal incentives have resulted in a dramatic slowdown in renewable energy initiatives.

Renewable energy use falls into three categories: thermal applications (passive solar building designs); fuels from biomass (firewood); and wind/solar electric. Hydroelectric power currently provides 6% of the worldwide energy consumption. The U.S. currently has the largest installed hydroelectric capacity in the world, providing up to 15% of electricity generation annually. Although hydropower production is dependent on annual rainfall, it offsets the need for about three times as much energy from fossil fuels because conversion efficiencies for fossil fuel plants are much lower.

Only 59% of hydropower potential in the U.S. was tapped in 1980; however, it is unlikely that capacity will increase in the near future. The major reason is that construction of dams for use as hydroelectric facilities can be extremely damaging to the environment. New technologies for smaller scale hydroelectric generators are being tested, however environmental problems again are expected to limit growth potential in this area as well.

Solar thermal energy and wind power generally have been considered to be the most promising of the newer renewable technologies. Wind farms saved the equivalent of about 1.5 million barrels of oil in 1985. The energy shortages of the '70s inspired a wave of wind machine construction. Federal tax credits spurred investment in the technology, as well as in solar installations. These incentives, however, have diminished and are set to expire this year. Together with the return of low oil prices, wind and solar energy technologies no longer are competitive with traditional energy sources.

Among liquid fuel substitutes for transportation applications, the most commonly recognized alcohol fuel is ethanol. It can be produced from a wide variety of feedstocks. Brazil has led the way in experimenting with ethanol as an automotive fuel, using its large feedstock of sugar cane. In the U.S., ethanol has been supported by farmers, who see it as a way to stimulate the farm economy and as a means to eliminate farm surpluses. However, ethanol production costs will not allow it to compete with gasoline currently (or methanol in the future) as a fuel supply extender. Ethanol's niche in the motor fuels market in the U.S. is most likely to be as an octane enhancer, and it should be competitive in this role by the end of the 1990s. Ethanol is not likely to compete well with methanol as a major supply source for motor fuel due to methanol's substantial and continuing cost advantage.

There are two major conclusions from this chapter's review of renewable energy sources. First, none of the non-fossil forms of electricity generation will provide an early alternative to continued, and, quite likely, increasing use of fossil fuels to meet electric power growth. Second, there are at present no liquid fuel substitutes for the transportation sector that approach the economics of petroleum products.

Chapter Three: Vital Statistics

Nuclear

The 60-megawatt reactor at Shippingport, PA, completed in 1957, was the first commercial reactor to generate electricity for the power grid.

Of 249 nuclear plants ordered since 1953, 119 have been cancelled.

At the end of 1988, the U.S. had 109 reactors licensed to operate, with a generating capacity of 97.2 MWe.

There are nuclear plants operating in 33 states.

Residential electricity prices *tripled* between 1973 and 1984. Industrial electricity prices *quadrupled* over the same period.

Nuclear plant construction times have grown from 4 to 6 years in the 1960s to more than 15 years in the 1980s.

Nuclear plant nonfuel operating costs grew at 12% per year between 1974 and 1984. Comparable coal plant operating costs grew at only 2% per year.

The high-level nuclear waste repository is not scheduled to open until 2003.

The nuclear waste repository is required to protect the environment from radiation for 10,000 years.

The likelihood of any new nuclear plant construction by the end of the century, or even a new plant order, is very small.

Renewables

Energy efficiency in the U.S. in the 1970s was the lowest among the developed nations.

Energy demand in the U.S. was projected to exceed 100 quads by 1990; however, demand peaked at 79 quads in 1979.

Renewable energy sources, including hydro, provide 9 to 10% of U.S. energy supply.

Hydroelectric power provides 14 to 15% of U.S. electricity.

The U.S. had 16,769 wind turbines (1,463 MWe) by the end of 1987.

Current installed geothermal electric capacity is about 2,000 MWe.

Ethanol production in 1988 will use nearly 400 million bushels of corn.

Ethanol has an octane rating of about 119 and performs well as an octane enhancer, raising octane ratings by about 3 points in a 10% ethanol/90% gasoline mixture.

The price of ethanol is likely to remain substantially higher than that of gasoline or methanol on an energy content basis for the foreseeable future.

Chapter Three: End Notes

[1]I.C. Bupp, "Nuclear Power: The Promise Melts Away," in *Energy Future: A Report of the Energy Project at the Harvard Business School* Third Edition (New York: Vintage Books, 1983), p. 140.

[2]R. J. Mattson, testimony before the North Carolina Utilities Commission, "Harris Project Panel IV," Exhibit 2, Docket No. E-2, Sub. 537, April 27, 1988, pp. 7-10.

[3]U.S. Department of Energy, *United States Energy Policy 1980-1988*, (Washington: USDOE/S, October 1988), pp. 140-142.

[4]Deudney, Flavin, *op. cit.*, p. 27.

[5]*U.S. Energy Policy, op. cit.*, pp. 91-93.

[6]*Monthly Energy Review, op. cit.*, p. 80.

[7]*U.S. Energy Policy, op. cit.*

[8]*Ibid.*

[9]*Energy Security, op. cit.*, p. 190.

[10]*Monthly Energy Review, op. cit.*, p. 105.

[11]*U.S. Energy Policy, op. cit.*, pp. 96-97.

[12]Drawn from the transcript of a speech by the Honorable James R. Schlesinger before the Nuclear Energy Forum, Washington, D.C., October 31, 1988.

[13]U.S. DOE-Energy Information Administration, *An Analysis of Nuclear Power Plant Operating Costs*, (Washington: USGPO, March 1988), pp. vi-x.

[14]Christopher Flavin, "Reassessing Nuclear Power: The Fallout From Chernobyl," *Worldwatch Paper 75*, March 1987, pp. 40-41.

[15]George M. Bohigian, "Medical Perspective on Nuclear Power," Report of the Council on Scientific Affairs, American Medical Association, CSA Report G (I-88), January 1989, pp. 36-37.

[16]Flavin, *op. cit.*, p. 51.

[17]D'Alessandro, Bill, "Dark Days for Solar," *Sierra*, July-August 1987, p. 37.

[18]Best, Don, "Solar Cells: Still A Tough Sell," *Sierra*, May-June 1988, p. 28.

[19]Blackburn John O., *The Renewable Energy Alternative: How the United States Can Prosper Without Nuclear Energy or Coal*, 1987, p. 51.

[20]Brown, Lester R., *State of the World*, 1988, p. 63.

[21]Blackburn, 1987, p. 48.

[22]Brown, *op. cit.*, p. 64.

[23]*Energy Security, op. cit.*, p. 206.

[24]Modesto A. Maidique, "Solar Future," in Stobaugh & Yergin, *op. cit.*, p. 261.

[25]*Monthly Energy Review, op. cit.,* pp. 9-11.
[26]Wells, Ken, "Clouded Outlook: As A National Goal, Renewable Energy Has An Uncertain Future," *Wall Street Journal,* February 13, 1986, p. 1.
[27]Brown, *op. cit.,* p. 76.
[28]*Ibid.,* p. 76.
[29]Strand, Robert, "High oil prices, tax breaks fueled windmill industry," *Miami Herald,* 1/25/87, p. 3C.
[30]D'Alessandro, *op. cit.,* p. 34.
[31]*Ibid.,* p. 35.
[32]Brown, *op. cit.,* p. 31-32.
[33]Melloan, George, "Californians Will Pay Dearly for PURPA Power "*Wall Street Journal,* March 31, 1987, p. 37.
[34]Strand, *Miami Herald,* January 25, 1987, p. 3C.
[35]D'Alessandro, *op. cit.,* p. 34.
[36]*Ibid.*
[37]*Energy Security, op. cit.,* pp. 202-204.
[38]"Gasoline Finds Renewed Life as Substitute for Leaded Fuel." *PB Post Evening Times,* October 17, 1985, p. A18.
[39]Department of Energy, "Assessment of Costs and Benefits of Flexible and Alternative Fuel Use in the U.S. Transportation Sector," January 1988, p. D-6.
[40]Presentation of Eric Vaughn, President and CEO of the Renewable Fuels Association, 1988 Conference on Oxygenated Fuels, June 20-21, 1988.
[41]Owen, David, "Octane and Knock," *The Atlantic Monthly,* August 1987, p. 56.
[42]Vaughn, *op. cit.*
[43]G. Alex Mills and E. Eugene Ecklund, "Alcohols as Components of Transportation Fuels," *Annual Review of Energy* (1987), p. 48.
[44]Segal, Migdon R., *et al.,* "Analysis of the Possible Effects of HR 2031, Legislation Mandating the Use of Ethanol and Methanol in Gasoline," CRS Report for Congress, No. 88-71 SPR, November 17, 1987, pp. 27-31.
[45]Ibid., p. 80.

CHAPTER FOUR

SMOGLESS TRANSPORTATION FUELS

"Contrary to popular opinion, the main cause of the current oil price collapse is not increased fuel efficiency of vehicles. Instead, it is due primarily to the adoption in less than a decade of substitutes for oil by electric utilities, industry, other businesses, and households. Because transportation is using an increasing share of oil, the ability to implement substitute transportation fuels will be far more important when the next oil price run-up begins, possibly as early as the 1990s. History shows that substitution in transportation fuel takes decades, so the U.S. cannot risk waiting for another price run-up to implement petroleum substitutes in transportation."

-D.J. Santini

In the preceding chapters, we examined the relative merits of our traditional energy sources, and found that none had the capacity to replace petroleum completely in the transportation sector without sweeping changes in fuel supply or automobile design. In this chapter, we will examine a fuel that can be produced using any and all of the traditional feedstocks, as well as some very untraditional feedstocks, and could fuel the nation's automobiles with minimal modifications to existing automobile engines. The fuel is methanol, an alcohol that can be produced from coal, natural gas, or biomass and that burns more cleanly and efficiently than gasoline. In the following pages we will discuss the potential for methanol in the U.S. transportation sector: its properties as a neat (unblended) automotive fuel, its usefulness as a blending agent for gasoline, and its application as diesel fuel.

Methanol, methyl alcohol or wood alcohol, is a colorless liquid with molecular formula, CH_3OH. For over two hundred years it has been used as an energy source. Early processes distilled wood alcohol from a liquid that was produced during the manufacture of charcoal. In this process, one ton of wood produced about three to six gallons of methanol. As early as 1830 it was used for lighting purposes, but soon was replaced by whale oil; and its uses thereafter were limited to cooking and heating.[1] Methanol was the source of fuel used for early automobiles; but since it wasn't available commercially until two decades after gasoline was established in the transportation market, it was never used widely as an automotive fuel.[2]

Today methanol is used mainly for chemical purposes, in the production of formaldehyde and acetic acid. However, methanol has superior physical, chemical and combustion properties that make it an excellent automotive fuel. These properties are the reason methanol has been the official fuel of the Indianapolis 500 since 1965.

Methanol as an Automotive Fuel

Fuel Properties

Methanol's physical and chemical properties can be summarized as follows:

> *Methanol is a colorless liquid with a freezing point of -97.8°C and a boiling point of 64.5°C. Its specific gravity of 0.792 (20°C) translates into 6.59 lb/ gallon or 334 gal/MT. Methanol is miscible with water and virtually insoluble in crude oil and most nonaromatic hydrocarbons. Its heat of vaporization (8.98 kcal/mole) is greater than that of hydrocarbons of comparable boiling point but less than that of water. It will form explosive mixtures with air in the concentration range of 6.7% to 36.5% by volume.[3]*

While the properties of methanol as listed above may give little insight into its suitability as an automotive fuel, the differences become evident when it is compared to gasoline and other potential gasoline substitutes (see Table 4-1). As an alcohol, methanol possesses different physical, chemical and combustion characteristics which differentiate it from petroleum and necessitate certain modifications in its application as a transportation fuel. Methanol has a molecular weight that is about half that of gasoline, and an energy density that is 48% that of gasoline in the liquid form. Liquids, however, do not burn; and when the vapor form is considered, a much different picture emerges--a picture of a superior automotive fuel.

Table 4-1. Physical Properties of Selected Alternative Fuels, Gasoline, and Diesel Fuel

Properties	Unleaded Gasoline	Diesel Fuel (No. 2)	LPG (HD-5)	CNG	Methanol	Ethanol
Constituents	Mixture of hydrocarbons (chiefly C_4–C_{10})	Mixture of hydrocarbons (chiefly C_{12}–C_{20})	95% propane, 5% butanes	60-98% methane; remainder, ethane and other paraffins, CO_2, H_2, He, N_2	CH_3OH	C_2H_5OH
Boiling Range (°F @ 1 atm)	80 to 420	320 to 720	-44 to 31	-259[a]	149	173
Density (lb/ft³)	43 to 49	49 to 55	31[b]	8[c]	49.2	49.2
(lb/gal)	5.8 to 6.5	6.5 to 7.3	4.1[b]	1.07[b]	6.6	6.6
Energy Content (net)						
BTU/lb	18,700-19,100	18,900	19,800	21,300[a]	8,600	11,500
BTU/gal	112,000-121,000	123,000-128,000	82,000	22,800[a]	56,560	75,670
Autoignition Temp. (°F)	450 to 900	400 to 500	920 to 1,020	1,350	878	795
Flashpoint (°F)	-45	125 (min)	-100 to -150	-300	52	70
Octane Number Range (R+M) 2	87 to 93	N/A	104[d]	120[d]	99	100
Flammability Limits (vol% in air)[e]	L = 1.4 H = 7.6	L = 0.7 H = 5.0	L = 2.4 H = 9.6	L = 5.3 H = 14.0	L = 6.7 H = 36.0	L = 4.3 H = 19.0
Sulfur Content (wt%)	0.020 to 0.045	0.20 to 0.25	Negligible[e]	Negligible[e]	None	None
Flame Speed (ft/sec)	1.3	1.3	1.3	1.1	1.3	1.3

[a] Pure methane. Other minor constituents (ethane, propane, etc.) boil at higher temperatures.
[b] At 80 °F with respect to water at 60 °F.
[c] 100% methane, 80 °F at 2,400 psi with respect to water at 60 °F.
[d] Octane ratings above 100 are correlated with given concentration of tetraethyl lead in isooctane.
[e] L = lower; H = higher.
[f] Natural sulfur content is very low, though measurable.

Source: U.S. DOE Report, Assessment of Costs and Benefits of Flexible and Alternative Fuel Use in the U.S. Transportation Sector, January 1988.

The heat of vaporization of methanol is 3.8 times greater than iso-octane (a molecule that is representative of gasoline), so methanol absorbs a much greater amount of heat in changing state from a liquid to a gas. This characteristic gives methanol a much higher thermal efficiency in combustion. In fact, methanol *vapor* mixed with air has an energy density that is 13% greater than iso-octane vapor-air mixtures. This suggests that an engine designed to take advantage of methanol's unique characteristics could achieve fuel economy superiority over gasoline. To date, most methanol applications have been conversions of gasoline engines and have fallen short of gasoline fuel efficiency.

Simple molecular structure is methanol's most striking difference from gasoline. CH_3OH is a simple bond of one carbon, one oxygen and four hydrogen atoms. Hydrogen is universally acknowledged as the ultimate fuel because of its limitless supply, high energy content and pollutant-free emissions. Carbon is the least desirable element in a fuel, as it takes a great deal of energy to oxidize, it rarely is completely consumed, and residual carbon can combine to form undesirable compounds. Oxygen reacts with unburned particulates to form carbon monoxide and carbon dioxide gases. Some of the more complex hydrocarbons, such as coal and crude oil, also contain elemental sulfur and nitrogen that, in the combustion process, can lead to the formation of sulfur dioxides and nitrogen oxides. Methanol has fewer problems in all of these areas.

The nature of methanol's composition, with just one carbon atom to four hydrogen atoms, is more favorable to combustion. Because it takes energy to get rid of carbon, fewer carbon atoms means less energy is used to break the hydrogen-carbon bonds, leaving a greater portion to actually generate heat that can do work. This composition gives methanol a leaner flammability limit than gasoline, which means that it can be ignited in a lean (lower fuel-to-air) mixture without the cycle-to-cycle problems found in gasoline engines, thereby improving efficiency.

Methanol's polar molecular structure makes its bond very strong. While gasoline has a molecular formula that is much more complicated than methanol (crude oil is $C_{85}H_{172}$), the complexity of its bond makes it much weaker. The simplicity of methanol's bond makes it stronger, less volatile, and less likely to evaporate. This is a significant air quality gain, since escaping gasoline fumes (volatile organic compounds) are a major contributor to the production of ozone and smog. The Environmental Protection Agency has estimated that for every 1,000 gallons of fuel pumped into cars, nearly ten pounds of gasoline fumes are released into

SOURCES OF ELECTRIC ENERGY

PHOTOVOLTAICS

HYDROELECTRIC

WIND GENERATED

NUCLEAR POWER

FUTURE SOURCES

O.T.E.C.
GEOTHERMAL
OCEAN CURRENTS
WAVE MOTION
LOW COST FUSION

CONVERSION OF ENERGY TO USABLE FORMS

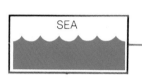

**REVERSE OSMOSIS
SYSTEM**
PURIFIES SEAWATER FOR USE IN
HYDROGEN GENERATOR.

HYDROGEN SYSTEM
GENERATES HYDROGEN FROM
FRESHWATER FOR METHANOL
PRODUCTION.

**SEAFUEL® SYNTHESIZER
PLANT**
COMBINES CO_2 AND H_2 TO
PRODUCE METHANOL

SEAFUEL® STORAGE
CONTAINERS OF METHANOL
PRODUCED FROM SEAWATER.

METHANOL APPLICATIONS

METHANOL PRODUCED FROM WINDMILL ELECTRICITY AND SEAFUEL® SYNTHESIS FUELS A PERKINS DIESEL ENGINE.

CLEAR EXHAUST PRODUCES MINIMAL ENVIRONMENTAL POLLUTION.

DUEL-FUELED CART USES GASOLINE OR SEAFUEL®.

DIESEL-METHANOL BUS OPERATING IN CALIFORNIA.

PRIVATE BOAT IN THE BAHAMAS USING SEAFUEL®.

METHANOL-FUELED TRUCK

THE STIRLING ENGINE

THE STIRLING ENGINE
CAN USE ANY FUEL INCLUDING
METHANOL

METHANOL-FUELED VEHICLES

Toyota (Hi-Lux Surf Van)

Honda (Accord)

Mitsubishi (Gallant Sigma)

Nissan (Caravan)

FUEL CELL

FUEL CELL
DIRECTLY CONVERTS HYDROGEN AND OXYGEN
TO ELECTRICITY.

PC-1401
TWO-MAN SUBMERSIBLE OUTFITTED WITH FUEL
CELL SYSTEM FOR EXTENDED MISSION DURATION.

HYDROGEN-FUEL APPLICATIONS

RECENTLY A HYDROGEN-FUELED AIRCRAFT HAS BEEN FLOWN IN RUSSIA NEAR MOSCOW BEFORE A GROUP OF INTERNATIONAL HYDROGEN EXPERTS.

A U.S. WORLD WAR II PILOT, WILLIAM CONRAD, HAS DEVELOPED AND FLOWN A SMALL AIRCRAFT USING HYDROGEN FUEL NEAR FORT LAUDERDALE, FLORIDA.

HYDROGEN-FUELED TRACTOR.

SPACE FLIGHT LIQUID HYDROGEN FUEL STORAGE TANK AT KENNEDY SPACE CENTER - HYDROGEN AND FUEL CELLS ARE THE PRINCIPAL SOURCE OF ONBOARD ENERGY FOR ALL SPACE FLIGHTS.

NASA PROPOSED HYDROGEN FUELED SUBSONIC JET TRANSPORT.

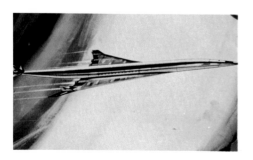

LOCKHEED PROPOSED HYDROGEN FUELED SUPERSONIC JET.

SEAFUEL SYNTHESIS PROCESS
SSP

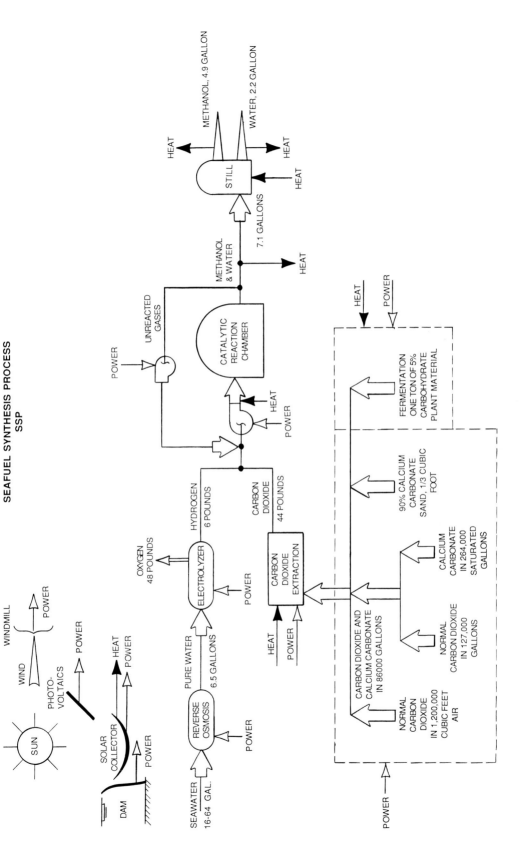

the atmosphere.[4] In addition to the refueling benefits, methanol's greater heat of vaporization makes it more difficult to ignite in open spaces, and therefore less likely to explode or burn in an accident. This is one reason racing car drivers prefer pure methanol as their fuel.

The octane rating of a fuel measures the speed at which it burns. Straight-run gasoline has an octane rating of about 80, far below the minimum requirement of 87. Low octane causes incomplete, inefficient combustion as the flame of the ignited fuel often burns too quickly, creating conditions that can cause the fuel-rich boundary layer in the cylinder to ignite prematurely, causing harmful detonations (engine knock) that interrupt the rhythm of the pistons and often causing engine damage.[5] With an octane rating of 120, methanol burns slowly, allowing a greater portion of its energy content to be utilized.[6] This increased thermal efficiency (greater than that of gasoline) means more of the fuel is burned before the cylinder exhaust valve opens, allowing fewer unburned hydrocarbons to escape.

Methanol's cooler burning substantially reduces the formation of nitrogen oxides, which are produced from the nitrogen and oxygen in the air when the temperature of the combustion process is high. The lower carbon content of the fuel also reduces the production of carbon monoxide, which lessens the need for a catalytic converter. Elimination or reduction of these controls would lower the cost of the vehicle and further improve mileage.

Methanol's higher octane and its lower flame temperature mean the cylinder can compress the fuel-air mixture into a smaller volume before it is ignited. Thus an engine designed for pure methanol can accommodate a higher compression ratio (11.5:1 or higher vs. 8-9:1), resulting in a methanol-filled cylinder pushing the piston harder.[7] A car gets about 25% more power from a methanol-burning high-compression engine versus a gasoline engine of equal size with a lower compression ratio.[8] This increase in power, coupled with better performance, results in about 10% better fuel economy, so that only about 1.8 gallons of methanol (rather than 2, as implied by energy content) are needed to replace a gallon of gasoline in conversions of gasoline engines.[9]

Ultimately an engine design suited specifically for the combustion of neat methanol would offer performance superior to that of a comparable gasoline engine, but currently a few problems remain that need to be considered. Methanol, like other chemically active substances is corrosive with some materials, most significantly zinc, lead, aluminum and magnesium; so a car designed specifically for its use would have to use compatible materials. It also reacts with certain elastomers used in rubber fuel lines and gaskets, causing them to swell or become brittle. This can be remedied by increasing the fluorine content of the elastomer.

Methanol does not evaporate as easily. As a result, a neat methanol engine will experience some trouble with cold starting below 50°F, unless the fuel is vaporized. This has been a problem for methanol-fueled vehicles on occasion, but usually can be remedied by mixing-in a small percentage of a more volatile hydrocarbon, such as propane. One alternative to this use of a blending agent is already being investigated in the use of a newly patented valve called a "thermocharger," which uses a fine double-screen to atomize the liquid fuel into small droplets that can rapidly absorb the necessary heat to change from liquid to vapor. These droplets have a much larger surface area, which greatly increases the rate of energy transfer. This approach offers some interesting opportunities for future efficiency gains in methanol engine design. At best, existing gasoline engines are 25% efficient, with the average automobile reaching only 15% efficiency when new and falling well below that level as it ages. In the conversion process, 75%-85% of the heat energy in the fuel is lost. Because the efficiency is so high with fully vaporized methanol, it may soon be possible to achieve a 1 to 1 fuel equivalence or greater. Tests done by GTE and The Bank of America on a Chevrolet Citation fitted with the valve proved that methanol fuel economy (19.3 mpg) can nearly equal that of a gasoline-fueled car of the same model (20.7 mpg).[10]

One other problem facing neat methanol engines is abnormal wear on the engine during the warm-up phase of operation. This results from the presence of liquid methanol, which collects on the cylinder walls and dilutes the engine oil, reducing its lubricity. Vaporizing the methanol fuel solves this problem. This can be remedied with appropriate crankcase lubricating oils, and several oil companies have made progress in the search for these.

Health and Safety

The hazards associated with methanol are comparable to those associated with gasoline, and findings on the health and safety aspects of alcohol fuels are generally more favorable toward use of alcohols than gasoline.[11] The types of hazards associated with methanol are those resulting from human contact, combustion, and dangers it could pose to the environment.

Methanol is not highly toxic (it is less toxic than gasoline), but 30 to 100cc can be lethal if ingested.[12] Absorption of large amounts into the bloodstream can produce weakness, heat sensation, abdominal pain, vomiting, blurred vision, temporary or permanent blindness and convulsions. In extreme cases it can cause death. Hundreds of people die each year from accidents involving syphoning of gasoline, and this will probably be the greatest risk associated with methanol's toxicity.[13] While the unhealthy effects of inhalation of methanol vapors are comparable to those of gasoline, it must be noted that methanol is less volatile than gasoline, meaning that it evaporates less easily, and is therefore less readily inhaled from normal exposure. Exposure to gasoline through ingestion, inhalation or skin contact is considered more irritative and dangerous to health than methanol.[14]

The major safety hazards associated with both methanol and gasoline are fire and explosion. Methanol has a higher flash point than that of gasoline and has a higher electrical conductivity than gasoline, making it more difficult to ignite. Thus in open air, gasoline is more hazardous than methanol. However, in confined areas, such as the fuel tank, the situation is the reverse. In a gasoline fuel system, the vapor space in the storage tank is too rich to burn. The oxygen present in methanol, however, makes it combustible. This problem has been addressed in several ways, such as "bladder" fuel tanks, flame-arresting foams, or adding light hydrocarbons that enrich the vapor space over the liquid in a closed container above the upper explosive limit.

Adding hydrocarbons to methanol can solve other problems as well. The discussion above already has addressed the need for a volatile additive to solve cold starting problems, which may be remedied by a shot of propane. Hydrocarbon additives also give luminosity to the nearly invisible flames of neat methanol. Flame visibility in bright sunlight is achieved with 15% volume gasoline.[15]

If a methanol fire does start, there are some advantages over gasoline fires. Burning methanol radiates less heat than burning gasoline. Furthermore, fighting alcohol fires is easier than fighting gasoline fires because of the solubility of alcohols in water. A fogging nozzle should be used to prevent spreading the fire however, because methanol concentrations as low as 21% are still flammable.[16]

Environmental Aspects of Methanol Combustion

With regard to effects on air quality, methanol is widely acknowledged as environmentally preferable to gasoline; and a more detailed account of the effects of methanol combustion on air quality will follow. The reduction in nitrogen oxide emissions from neat methanol

has already been mentioned.   Unburned fuel emissions are comparable
to gasoline use, but constituents are not as reactive.   One concern is
formaldehyde emissions, which are produced in greater quantities by
alcohol-fuel engines and are the dominant aldehyde in motor vehicle
exhaust gases.   Formaldehyde is an irritant to the eye, nose, throat and
upper respiratory tract.   It is unstable in air and decomposes rapidly so
that its effects are localized.   Although there is some controversy over the
carcinogenic properties of formaldehyde, the National Cancer Institute
found little evidence that it causes cancer in the occupational work forces
exposed to it.   The low levels of formaldehyde produced by methanol-
burning vehicles are not considered hazardous at this time; and while
there have not been definitive studies on the effects of formaldehyde, the
moderate increase that would result from wide use of alcohol fuels is not
considered an elevated hazard.   Aldehyde emissions from methanol
combustion are only one-tenth the volume of total hydrocarbon emissions
and they degrade much more rapidly.[17]

     While some critics point to the limited amount of documentation
concerning the effects of prolonged exposure to formaldehyde, the
deleterious effects of gasoline fumes are well-documented. An American
Petroleum Institute study in 1984 found that rats and mice developed
cancer from breathing gasoline vapors. The EPA concluded that gasoline
vapors would be likely to cause human cancer as well.[18]

     The combustion properties that make methanol a superior
automotive fuel signify benefits for the environment as well. In July 1987
the EPA's Office of Mobile Sources published a paper entitled, "Air
Quality Benefits of Alternate Fuels," which projected the emissions
reduction potential on a per-vehicle basis for various concentrations of
methanol fuel.   The agency examined three types of vehicles capable of
burning methanol. Two types of vehicles currently being tested burned a
blend of 85% methanol, 15% gasoline, referred to as M85.   A third,
"advance technology" vehicle was designed to maximize the benefits of
pure, "neat" methanol, or M100, as an automotive fuel.   Test results
indicate that "neat" methanol vehicles could achieve reductions of 85 to
90% in volatile organic compounds emissions while reducing carbon
monoxide emissions 30 to 90% (both VOCs and CO emissions contribute
to ozone production). A vehicle burning a combination of 85% methanol
and 15% gasoline (M85) would reduce VOC emissions 20% to 50%.
Both vehicles would also comply with stringent new NO_x emissions
standards.   While methanol's low flame temperature does offer lower

NO_x emissions with the proper equipment, the EPA believes auto manufacturers will likely trade-off methanol's low NO_x characteristics to gain other benefits, such as increased fuel economy and performance, or a less expensive catalytic converter.[19]

Another area of environmental advantage is in potential soil and water pollution. The effects of methanol spills will be markedly different than gasoline spills because of methanol's solubility with water. Surface spills should pose no problem because methanol mixes easily with water, is quickly diluted to non-toxic levels, leaves no residue, and is easily decomposed by anerobic bacteria. Furthermore, effects of methanol on underground water are regarded as less damaging than those of gasoline and oil. Microorganisms in close proximity to a methanol spill will be greatly affected. However, unlike an oil spill, where the harm is concentrated in the area of the spill and organisms affected may take several months to recover, methanol's solubility with water will disperse the effects over a greater distance, lessening the damage. With methanol, bacteria and fungi return to the immediate area of the spill in about three weeks; and higher organisms return shortly thereafter.[20]

Methanol Blends

Gasoline Mixtures

Blending methanol with gasoline can both capitalize on some of methanol's desirable qualities as an automotive fuel and extend gasoline supplies. Concentrations of as much as 10 to 15% methanol can be blended with gasoline before changes in carburetion are necessary. However, alcohol-gasoline blends can lead to difficulties that are not found with either pure alcohol or straight gasoline use.

The benefits of alcohol-gasoline blends can be attributed partly to methanol's leaner burning characteristics. The most immediate result is a leaner burning engine, which produces lower hydrocarbons and carbon monoxide emissions. Alcohol blends (ethanol, methanol with 3.7% oxygen content) can achieve greater than a 22% reduction in CO emissions from older vehicles.[21] However, leaner fuel mixtures burn in excess oxygen; and this could cause an increase in nitrogen oxide emissions if it were not for the fact that the combustion temperature is lowered by the cooler-burning methanol. A second benefit of leaner burning, however, is increased thermal efficiency, meaning that fuel economy can be higher than expected. Some data have even shown better fuel economy both on an energy (per BTU) basis and on a miles per gallon basis as well.[22]

The two major problems with alcohol fuel mixtures are vapor lock and phase separation. Vapor lock occurs because the vapor pressure of the alcohol-gasoline blend is greater than the vapor pressure of either gasoline or methanol alone. Evaporative emissions increase as a result; and occasionally, vapor can form at the intake of the fuel pump (where pressure is lowest), often obstructing the fuel line.[23]

Phase separation occurs in the presence of even small amounts of water. While methanol is soluble with water, gasoline is not. When water is present, methanol and gasoline will separate; and the methanol, along with the water and the aromatic components of gasoline (benzene, toluene) will settle to the bottom of the fuel tank because their solution is denser. A car that is suited for gasoline will be unable to use the concentrated methanol mixture at the bottom of the tank.[24] Remedies include: increasing the concentration of alcohol in the fuel; increasing the gasoline's aromaticity; increasing the temperature of the fuel; or adding co-solvents such as ethanol.[25]

MTBE

Another blending agent that is growing in popularity is methyl tertiary butyl ether (MTBE). Made by reacting isobutane with methanol, MTBE enjoys some of the benefits of alcohol fuels and a greater compatibility with gasoline. Furthermore, because it can be made in the oil refinery, it offers the gasoline refiner an octane enhancer that requires little or no capital investment. It is also less expensive than ethyl alcohol (ethanol). Though not an alcohol, MTBE in gasoline offers to a lesser extent the benefits of an oxygenated blend. Carbon monoxide reductions of 14% have resulted from its use in Denver's oxygenated fuels program, as opposed to 30 to 40% projected from ethanol use during the same period.[26] MTBE is less volatile than gasoline-alcohol blends, meaning that its noxious fumes are less likely to evaporate; this, along with the hidden costs of ethanol have led some cities to prefer MTBE as an additive.[27] MTBE, however, unlike ethanol or methanol, could never be used by itself as an automotive fuel. It will remain a blending agent for other fuels.

Diesel Fuel Blends

While most individuals drive gasoline-fueled vehicles, diesel engines provide power for buses, trucks, marine engines and many industrial applications. Diesel engines have several aspects that are different from gasoline engines. While a typical gasoline-powered engine has a compression ratio of about 8:1, a diesel engine has a 17:1 compression ratio. (A methanol engine would be approximately 12:1). Rather than using spark plugs to ignite the fuel, the diesel engine relies on compressing the air in the cylinder until its temperature exceeds the point needed to ignite the fuel. As the piston begins the downstroke, fuel is sprayed into the hot chamber and immediately begins to burn. A certain level of fuel volatility is needed for diesel engines to function properly, and the ease with which diesel fuel will ignite is measured by its "cetane" number, or the percentage of methylnapthalene (cetane) in the fuel. Diesel has a cetane rating of 40 to 45, while methanol's tight chemical bond results in the low cetane number of 3, hardly conducive to auto-ignition without an additive.[28]

Diesel became more popular in the automobile market during the late 1970s when, following the oil crisis, the Energy Policy and Conservation Act of 1975 set its first Federal Fuel Economy Standards. Because diesel-powered cars boasted a higher fuel economy than gasoline powered cars, and diesel fuel price was lower than gasoline at the time, their fuel cost savings approached 40% for compacts and 25% for standard size cars.[29]

Although diesel fueled engines offer benefits of fuel economy and lower fuel costs, they are not without their drawbacks, most significant of which is their high levels of nitrogen oxide and particulate emissions. Heavy duty diesel engines are a major source of pollutants in urban areas. Transit buses are responsible for as much as 50% of total diesel particulate loading in some central city areas.[30] In March of 1985, the Clean Air Act mandated emissions reductions from diesel engines, beginning in 1991 for buses and in 1994 for heavy duty trucks.[31]

While the means of compliance are left to the discretion of bus manufacturers, research to date has not yielded a method of reducing particulate emissions on large truck and bus applications; and the EPA does not believe nitrogen oxide reductions are feasible.[32] For this reason methanol has been examined as a replacement fuel for diesels. The substitution of methanol in diesel engines has produced some impressive environmental results: sulfur emissions are eliminated; nitrogen oxides are reduced 80%; carbon monoxide emissions are lowered by 70%; particulates are eliminated or greatly reduced; reactive hydrocarbon emissions are negligible; and there is less objectionable odor.[133] In addition to these emissions reductions, methanol offers improved efficiency, and better acceleration.[34]

Although low auto-ignition characteristics make methanol an unattractive diesel fuel in its neat form, the potential benefits to the environment have spawned research efforts to solve problems of ignition. Several technical approaches have been attempted. These include changing the concentration of methanol, experimenting with different methods for igniting the methanol, and adapting engine design.[35] By injecting the fuel over a period of time, for example, some of the fuel may be able to burn while the rest is being injected. By controlling the cylinder temperature in the two-stroke engine, Detroit Diesel Allison (DDA) was able to achieve auto-ignition of methanol in spite of its low cetane number.[36] In early 1983, DDA began development of the world's first two-cycle compression ignition methanol engine; and in January 1984, the first methanol-fueled bus went into operation in San Francisco's Golden Gate District. According to DDA, the engine's power and performance is equal to or better than standard diesels. Fuel economy is comparable to diesel on a BTU content basis, and the exhaust is the cleanest of any presently available compression ignition engine.[37]

Other neat or near-neat approaches to alcohol use in diesel engines include active ignition systems (spark or glow plug), dual-fuel (using a diesel pilot and methanol for the power) and chemically spiked alcohol (cetane enhancers are added). Perhaps the most successful of these has been the ignition additive called "Avocet", a cetane enhancing additive and corrosion inhibitor containing lubricity additives and produced by Imperial Chemical Industries of the United Kingdom. During the period 1984-1987, the Aucklund Regional Authority in New Zealand operated two MAN SL200 standard city buses on neat methanol with the Avocet additive and achieved satisfactory operation from one of the buses.[38] Canada is also interested in methanol-fueled buses, and Canada's Mines and Resources Department is sponsoring a program to bring in as many as 20 methanol buses.[39]

The introduction of methanol into diesel applications holds significance beyond its environmental benefit. Because fleets of urban buses are fueled at a central location, there is not the usual need for widespread availability of refueling stations. Furthermore, if heavy-duty engine manufacturers decide to use methanol engines for urban transit buses as a means of complying with emissions regulations, the availability of a methanol engine will result without government intervention, thus making a first step in resolving the chicken and egg dilemma of fuel supply and availability of methanol burning vehicles.

Chapter Four: Summary

Methanol has been used as a source of energy for over two hundred years and was the source of fuel for early automobiles. While its lack of commercial availability prevented its wide-spread adoption as an automotive fuel, as a neat fuel it is actually superior to gasoline; and for this reason it is the standard fuel of the Indianapolis 500.

Methanol's simple structure, and relatively higher hydrogen to carbon content account for physical, chemical and combustion characteristics more desirable than those of gasoline. An automobile designed to run on pure methanol could get up to 20% more power than a comparable gasoline vehicle and would benefit from an increase in efficiency that could give it equivalent mileage with gasoline.

Methanol can offer environmental benefits as well. Increased thermal efficiency reduces the production of nitrogen oxide emissions while a higher oxygen content, and fewer carbon molecules, reduce carbon monoxide emissions. Methanol's higher heat of vaporization also means it is less likely to evaporate, reducing the amount of volatile organic compounds that are released into the atmosphere.

The desirable properties of alcohols have led many to suggest the use of alcohols as blending agents in gasoline to take advantage of octane enhancing and oxygenation benefits of alcohols without the transition to alcohol fuels. Yet the very differences between alcohol and gasoline present some negatives that reduce the benefits of alcohols when blended with gasoline. However, significant air quality gains can and have been achieved through the use of alcohol blends. The most commonly used blends are ethanol and MTBE, but neither can be justified as a straight automotive fuel.

Methanol's use in the transportation sector is not limited to its replacement for gasoline applications, and recent research efforts have extended its potential to the diesel fuel market. Several diesel buses nationwide are currently operating on methanol, eliminating the production of sulfur dioxide and diesel particulates and greatly reducing nitrogen oxide emissions.

Chapter Four: Vital Statistics

One ton wood produces: 3-6 gallons methanol via dry distillation

Number of years methanol has been used: over 200

Energy content of methanol: 56,600 BTU/gallon

Energy content of gasoline: 115,400 BTU/gallon (average)

Methanol chemical formula: CH_3OH

Chemical formula of ethanol: C_2H_5OH

Crude oil chemical formula: $C_{85}H_{172}$

Octane number of unleaded gasoline: 82-100

Octane number of methanol: 88-115

Octane number of ethanol: 100

Octane boost of 10% ethanol when added to 90% gasoline: 3 points

Thermal efficiency of methanol versus gasoline: 6% higher

Power of methanol burning engine versus gasoline engine of comparable size: 10% higher

Fuel economy of methanol burning engine versus gasoline engine of comparable size: 10% higher

M85: 85% methanol, 15% gasoline mixture.

M100: 100% methanol (neat methanol).

Limits set by EPA on gasoline blends:

MTBE	11%
Ethanol	10%
Ethanol-methanol blends	7.5%

Percentage of particulate emissions caused by diesel buses in some central city areas: 50%

Reduction of NO_x emissions by substitution of methanol in diesel buses: 80%

Reduction of CO emissions: 70%

Chapter Four: End Notes

[1]Mills, Alex G. and E. Eugene Ecklund, "Alcohols As Components of Transportation Fuels," *Annual Review of Energy,* 1987, p. 49.

[2]Lemonick, Michael, "Kicking the Gasoline Habit," in *Science Digest,* May 1984, p. 29.

[3]Marsden, S.S. "Methanol as a Viable Energy Source in Today's World," *Annual Review of Energy,* (Palo Alto: Annual Reviews, Inc.), p. 334.

[4]Pasztor, Andy, "U.S. Wants to Combat Gasoline Fumes From Cars Refueling at Service Stations," *Wall Street Journal,* July 17, 1984.

[5]Owens, David, "Octane and Knock," *Atlantic Monthly,* August 1987, p. 54.

[6]Octane numbers vary from source to source, but methanol's octane is usually cited as between 88 and 115, while unleaded gasoline is usually rated between 82 and 100.

[7]Simonietti, Toni, "Methanol: Classic Fuel Fits the Future," *GM Today,* October 11, 1985, p. 7.

[8]Lemonick, *op. cit.,* High compression engines are heavier and more expensive, however, thus offsetting some of the mileage gain.

[9]Mills & Ecklund, *op.cit.,* p. 55.

[10]*Alcohol Week,* August 1, 1983, p. 2.

[11]Mills and Ecklund, *op. cit.,* p. 69.

[12]Lincoln, John Ware. *Methanol and Other Ways Around the Gas Pump,* (Charlotte: Garden Way, 1976), p. 23.

[13]Mills & Ecklund, *op. cit.,* p. 71.

[14]*Ibid.*

[15]*Ibid.,* p. 70.

[16]*Ibid.*

[17]Lincoln, John Ware, 1976, p. 26. Volkswagen investigators found that blends lowered emissions significantly. 100% methanol, when blended with small amounts of water, lowered NO_x emissions and achieved nearly a 40% reduction in aldehydes.

[18]Taylor, Robert E., "Oil, Auto Industries Square Off in an Attempt to Deflect Burden of Meeting Gas Fumes Plan," *Wall Street Journal,* June 30, 1986, p. 56.

[19]EPA, "Air Quality Benefits of Alternate Fuels," June 1987, p. 4.

[20]Mills & Ecklund, *op. cit.,* p. 71.

[21]Fraas, Arthur and Albert McGartland, "Alternate Fuels for Pollution Control: An Empirical Evaluation of Benefits and Costs," given at the Western Economic International Association, Alternate Fuels Session, Los Angeles, June 30 - July 3, 1988.

[22]Reed, T.B. and R.M. Lerner, "Methanol: A Versatile Fuel for

[22]Reed, T.B. and R.M. Lerner, "Methanol: A Versatile Fuel for Immediate Use," *Science,* December 28, 1973, p. 1299.

[23]Pasternak, Alan, "Methyl Alcohol - A Potential Fuel for Transportation," from *The Energy Technology Handbook,* 1977, pp. 4-47.

[24]*Ibid.*

[25]Mills and Ecklund, *op. cit.,* p. 52.

[26]"High MTBE Sales Lower Colorado's Carbon Monoxide Reduction Figures." *Alcohol Week,* February 1, 1987, p. 1.

[27]"EPA volatility regulations may favor MTBE over ethanol to replace butanes," *Alcohol Week,* 8/18/87, p. 1.

[28]Bernton, Hal, William Kovarik, and Skott Sklar. *The Forbidden Fuel: Power Alcohol in the 20th Century,* 1982, p. 167.

[29]Bernton, *op. cit.,* p. 23.

[30]Wilson, Richard D., Director of Office of Mobile Sources, Office of Air and Radiation, U.S. EPA before the Subcommittee on Fossil and Synthetic Fuels, November 20, 1985. p. 8.

[31]Simonietti, Tony, *GM Today,* October 11, 1985, p. 7.

[32]Cannon, Joseph, testimony before the Subcommittee on Fossil and Synthetic Fuels, April 4, 1984.

[33]California Energy Commission, Methanol One Fact Sheet

[34]*Ibid.*

[35]Mills and Ecklund, *op. cit.,* p. 56.

[36]Toepe, R.R., J.E. Bennethum, R.E. Heruth, "Development of Detroit Diesel Allison 6V-92TA Methanol Fueled Coach Engine," SAE Paper no. 831744.

[37]"Methanol fueled bus brings clean transit," *Design News,* December 19, 1983, p.15.

[38]Aucklund Regional Authority, A Report on New Zealand Bus Trials Using Ignition Improved Methanol, Aucklund, New Zealand July 1984-1987.

[39]Simonietti, T., *op. cit.,* p. 7.

CHAPTER FIVE

CRACKING THE CHICKEN AND EGG DILEMMA

The United States is better able than other countries to regain some mastery over events. We have the domestic resources, the technological skills, and the creative genius to transform the world's energy market. Therefore, decisive American efforts to implement an energy policy would transform the international environment -- reducing the cartel's stranglehold on the world economy, restoring world prosperity, dispelling the clouds of political demoralization, and strengthening our security.

-Henry Kissinger

In the preceding chapters we have discussed the need for an alternative transportation fuel and the potential benefits of methanol. In this chapter we will discuss the impediments to methanol's entry into the fuel market, the various policy initiatives and current legislation aimed at removing these impediments, and the economics of methanol's implementation as an automotive fuel.

While methanol is already produced in significant quantities as a chemical, its introduction as neat methanol into the transportation fuel market depends on both the production of methanol-fueled vehicles and the development of an adequate (and accessible) fuel supply. Even with competitive fuel prices, neither automobile manufacturers nor fuel suppliers are willing to take the first step to meet these needs without substantial government intervention. Auto makers are reluctant to mass produce vehicles without assurances that both adequate fuel supplies and a fuel distribution infrastructure will be in place to meet the demand. Producers and distributors, on the other hand, have no incentive to invest in the necessary technologies without the vehicles to burn it. Just as with the classic chicken and egg dilemma, the question poses itself, "Which comes first?"

Consequently, the entry into the fuel market has been limited to government test programs. Government demonstration programs allow both significant testing experience with methanol-fueled automobile fleets and a central refueling location that can supply the methanol. In a similar manner methanol bus demonstration programs provide a niche for heavier vehicle testing experience.

Federal Demonstration Programs

In the fall of 1984, Congress appropriated $980,000 to the Department of Energy for the implementation of a methanol demonstration program for federal vehicles.[1] The plan consisted of a two-phase approach with the ultimate intent of integrating at least 1,000 M85 (85% methanol, 15% gasoline) vehicles into government motor pools. In the first phase, a minimum fleet of 5 gasoline vehicles at each site were to be converted for state-of-the-art use of methanol, some with compression ratios adjusted upward for optimal methanol use, some with no adjustments in the compression ratio. These were to be paired with a similar number of gasoline vehicles (five minimum) used as a control group (see Table 5-1). A second phase has been proposed in which up to

Table 5-1

DOE METHANOL VEHICLE FLEET AS OF OCTOBER 1987

Location	Type of Vehicle	Methanol	Gasoline[*]
California:			
Lawrence Berkeley Laboratory	1984 Chevrolet Citation	5	5
Illinois:			
Argonne National Laboratory	1986 Ford Crown Victoria	5	4
	1986 Chevrolet S-10 pickup truck	5	5
Total		15	14

[*]The gasoline vehicles were used as control vehicles to provide a source of comparison with the methanol vehicles.

Source: Ecklund and McGill

1,000 vehicles that were designed and manufactured to burn methanol would be purchased by GSA and integrated into the Federal fleet. Phase I currently is in operation at three sites, and Phase II has not yet been initiated.[2]

In addition to the DOE program, the President's Task Force on Regulatory Relief has recommended that GSA issue a request for proposals to procure 5,000 flexible-fueled vehicles for the Federal fleet. This proposal, announced in July of 1987, has not yet been acted upon.

Methanol Fleet Operations

As of June 1987, 29 vehicles were operating in two fleets nationwide.[3] The Lawrence Berkeley Laboratory Test Fleet has operated 5 methanol-fueled (88% methanol, 12% gasoline) 1984 automatic transmission Chevrolet Citations with 2.8 liter V-6 engines alongside 5 gasoline powered cars of the same model since November 1985.[4] The fleet operates the route between Berkeley and San Francisco, with the methanol vehicles averaging about 36 miles per trip. According to fleet managers, perceptions have been positive toward the methanol vehicles, which have not needed any major repairs. Because the Bank of America in San Francisco operates its own methanol fleet, refueling is no problem, as there are fueling points at both the laboratory and in San Francisco. Difficulties in project start-up and maintenance were minimal, and by October 1987 the methanol vehicles had accumulated 37,000 miles.[5] The methanol fleet reported some reduction (10% lower efficiency) in fuel economy and somewhat higher engine wear. It is important to note, however, that these were gasoline powered automobiles that were retrofitted for methanol use, rather than vehicles designed for methanol. Emissions data on these vehicles showed carbon monoxide and nitrogen oxide levels to be reduced considerably when using methanol fuels, while other hydrocarbon emissions increase somewhat (Table 5-2).[6]

The fleet at Argonne National Laboratory in Chicago consists of ten 1986 Chevy S-10 pickup trucks (5 methanol powered, 5 gasoline powered) and nine 1986 Ford Crown Victorias (5 methanol, 4 gasoline). These vehicles are powered with 5.0 litre, sequentially timed, port fuel-injected V-8 engines; and all have automatic transmissions.[7] To date the pickup trucks have encountered only minor problems, while the Crown Victorias have experienced some difficulties, primarily with plugged fuel injectors.[8] However, as the Argonne fleet had been operating less than

Table 5-2

EMISSIONS AND FUEL ECONOMY OF FIVE
CITATIONS BEFORE AND AFTER CONVERSION

Vehicle ID (License #)	Emissions (g/mile)			Fuel Economy	
	CO	HC	NO_x	l./100 km	km/GJ
E-36753					
(g)[a]	2.23	0.39	0.79	12.3	252.1
(m)[b]	1.87	0.98	0.66	24.1	233.9
E-36754					
(g)	7.86	0.21	1.15	12.3	252.9
(m)	2.38	0.59	1.06	23.0	245.0
E-36755					
(g)	2.32	0.19	0.94	12.3	252.6
(m)	2.27	0.75	0.80	24.6	229.2
E-36756					
(g)	2.89	0.20	1.01	11.9	261.3
(m)	3.28	1.27	0.53	24.5	230.2
E-36757					
(g)	15.91	0.44	0.96	12.5	249.2
(m)	1.55	0.74	0.29	23.6	238.4
Five-car average					
(g)	6.24	0.36	0.97	12.3	253.6
(m)	2.27	0.87	0.67	23.9	235.3

[a]Unleaded gasoline (per FTP requirements).
[b]88% methanol plus 12% unleaded gasoline.

Source: Ecklund and McGill

one year at the time of this report, and it was thought to be too early to reach significant conclusions. The Department of Energy initially expected Phase I to be completed in FY (Fiscal Year) 1986 as a basis for the purchase in 1987 of the 5,000 vehicles as planned in Phase II. However, in July 1987 they announced they could not estimate when Phase I would be finished and were uncertain as to whether or not Phase II would be undertaken.[9]

While the Methanol Demonstration Program has been successful with regard to the operation of the methanol fueled vehicles, the effectiveness in increasing the use of methanol fuel and methanol-fueled vehicles has been significantly less than desired. Although the DOE (Department of Energy) sent letters to 20 agencies with large vehicle fleets, only the Department of Defense participated in the alternative fuels program. Reasons for lack of participation were given as follows in an account by the General Accounting Office:

- Budgetary constraints--the DOE wanted at least five methanol and five control vehicles; fleet managers didn't want so many.

- Fleet mangers did not want to give up gasoline-powered vehicles of known dependability (which could be refueled anywhere) for methanol-powered vehicles of lesser known dependability and which could only be refueled where methanol is available.

- Some fleet managers did not want the automobile models DOE chose for methanol demonstration.

- Some fleet management personnel didn't want to be burdened with the additional responsibility, required under the demonstration program, of familiarizing drivers with operating methanol vehicles, collecting operating and performance data, training fleet mechanics, installing and maintaining experimental fueling facilities.[10]

The experience with the Methanol Demonstration Program for Federal Vehicles holds a valuable lesson for the introduction of methanol vehicles into the commercial fleet. The Reagan Administration firmly opposed intervention in the marketplace, and instead encouraged the use of methanol in its own federally funded, centrally fueled fleet, theoretically circumventing both the problems of fuel supply and vehicle compatibility, and thus providing a point of entry for methanol into the fuel market. Yet the conclusion of the General Accounting Office report, "Alternative Fuels: Information on DOE's Methanol Vehicle Demonstration Program," was that the DOE Program has done little to encourage the increased use of methanol fuel and vehicles *within* the federal fleet nor has it promoted production of methanol vehicles and distribution of methanol fuel. The chicken and egg dilemma doesn't seem to be confined to the commercial market. So how can methanol's introduction be encouraged? The GAO reported, "Program managers agreed that without incentives to make the methanol vehicles more attractive for federal fleet managers, few would be willing to take the risks that come with the uncertainty."[11] If such is the level of receptiveness of the Federal government to innovation, how can methanol be expected to enter the transportation sector at a commercial level on market force alone?

Introducing Methanol to the Marketplace

The Fuel Approach--A Blends Strategy

Alcohol was initially blended in gasoline as an octane enhancer, a replacement for lead in the phaseout of that substance ordered by the EPA in 1973. More recently, alcohol blends have been used to increase the oxygen content of fuel in cities with unacceptable levels of ozone pollutants. While it is widely acknowledged that alcohol-gasoline blends lack the full benefits of neat alcohol (volatility increases and consequently so do hydro-carbon emissions) some see the increased use of blends as a point of entry into the fuel market for alcohol.

In the same year (1985) that the EPA issued its final rule on the lead phasedown, allowing no more than 0.1 gram of lead per gallon of gasoline by January 1986, the Agency granted DuPont a waiver from the Clean Air Act for a new blend of methanol and other alcohols with a maximum content of 6% methanol by volume and minimum co-solvents of 2.5%.[12] The DuPont waiver, as it is now referred to, has permitted an increase in methanol blending to about 280 million gallons per year, or about 0.09% of the demand of the U.S. transportation sector. About 95% of this, however, is used as a feedstock for MTBE, which is blended into about 1.2 billion gallons of gasoline per year.[13] Table 5-3 shows the cost-effectiveness of the three major alcohol-gasoline blending agents.

Current legislation mandating the use of alcohol blends is linked to the revision of the Clean Air Act. On June 12, 1989, President Bush announced his proposed revisions to the Clean Air Act.[14] The most far-reaching component of this proposal is the long-term clean fuels program. The program identifies nine major urban areas with the greatest concentration of ozone (Los Angeles, Houston, New York City, Milwaukee, Baltimore, Philadelphia, Greater Connecticut, San Diego, Chicago). In these areas, the program calls for a ten-year phase-in of alternative fuels and clean-fueled vehicle sales, starting with: 500,000 vehicles in 1995; 750,000 in 1996; and one million per year in 1997 through 2004.[15] This initiative provides a major and somewhat unexpected boost to support a more rapid increase in the use of alcohol fuels.

Proponents of the alcohol blends strategy cite as a primary advantage its slower, less disruptive, method of introducing methanol into the marketplace. Perhaps the most thorough argument for a blends policy can be found in a study commissioned by the U.S. Department of Energy Entitled, "Rationale for Converting the U.S. Transportation System to Methanol Fuel." The study argues that abrupt shifts in technology accelerate just before (and continue during) a several year period of slumping vehicle sales and often negative economic growth, and it correlates previous fuel and transportation technology transitions with periods of business depression in the U.S. Authors R.P. Larsen and D.J. Santini, researchers at Argonne National Laboratory in Illinois, argue:

Development and maintenance of a consensus in favor of a steadily expanding transition to methanol will be especially difficult for U.S. manufacturers and governments, given their proclivity to seek short term rewards, even at the expense of long term benefits . . . the research shows that they have historically attempted to implement fuel-switching transitions in transportation too late and too rapidly, sometimes with disastrous consequences.[16]

The examples they used include the creation of the railroads and the recession of 1836-46, the introduction of the steam engine and the recession of 1893-1898, and the invention of the internal combustion engine during 1927-35 and the Great Depression. Each of these

Table 5-3

COST EFFECTIVENESS OF OXYGENATED FUELS PROGRAM
IN 1990

I. Costs

	11% MTBE (2% Oxygen)	DuPont Waiver (5% Methanol) (2.5% Ethanol) (3.7% Oxygen)	10% Ethanol (3.7% Oxygen) Imports	Domestic
*Cost of Additive ($/gallon)	$.74	$.57	$.50-1.10	$(.27)-1.10
Cost of Gasoline ($/gallon)	.60	.60	.60	.60
Cost of Oxygen-enhanced Fuel	.61	.59	.59-.65	.45-.65

II. Emissions

	MTBE	DuPont Waiver	Ethanol
% Reduced	12	22	22
Ton Reduced per 4,000 gallons (High Altitude City)	.26	.47	.47

III. Cost per ton of Carbon Monoxide Reduced

MTBE	DuPont Waiver	Ethanol
$154	0	($1277) - $426

*All alcohol prices from "Alcohol Week", Vol. 8 No. 13, March 30, 1987.

Source: U.S. Department of Energy

transitions, they argue, involved a relatively abrupt increase in the thermodynamic efficiency of vehicle engines, and a switch to methanol would allow a similar shift.

The period from 1978-85 has witnessed a decrease in domestic automobile sales of over 40%, with only minor technological transition (lead phaseout, turbocharging and the adoption of fuel injection). While Larsen and Santini observed that, "the ultimate desirability of neat methanol seems obvious from a historical perspective," they caution both against waiting too long to implement methanol as an automotive fuel (as has been our mistake historically) and against going too far in the promotion of a fuel switch, (Brazil's mistake, in their opinion, in implementing ethanol).

In the short run, the demand for transportation fuel is very inelastic, meaning that a small increase in the supply of transportation fuels can drive the price of those fuels down dramatically. The effects of Saudi Arabia's increase in oil production illustrate this point dramatically. Therefore, it is desirable, from an oil company's point of view, to resist even small intrusions into its market. The consumers' and auto producers' interest, by contrast, is best served when intrusions into the market by competitors, even small ones, are easy.[17]

A blends policy would allow a slower, less disruptive introduction into the fuel market. Neat methanol vehicles would require an abrupt transition in automotive technology similar to those that have historically preceded economic depressions. In this way an engine designed to burn methanol blends would provide a transition point for the adoption of alcohol fuels.

While the smooth transition to methanol is unquestionably desirable, many of its supporters disagree that a blends policy is the best means to that end. The problem of volatility is perhaps the greatest objection. Because blends actually increase the volatility of the fuel (and hence hydrocarbon emissions) many believe this defeats the purpose of the environmentally preferable alcohol fuel.[18] Furthermore, while many of the performance problems associated with methanol blends can be solved with technology, the improper monitoring of blends cannot; and this may pose another problem. Finally, while the use of blends may provide a slower introduction of the amount of methanol entering the fuel market, it doesn't really remedy the problem of producing automobiles suited for neat methanol. For this reason, the Department of Energy rejects the use of blends as a transition to methanol as a transportation fuel. A September 1988 Report on Alcohol Fuels declared, "Methanol blends are not generally considered to be a stepping stone to neat methanol because of significant differences in conversion and transition costs between neat methanol and blends."[19]

Auto Industry Initiatives

Attempts to encourage methanol use in the production of compatible vehicles have ranged from appeals for mandatory purchase of a federal alcohol fleet to emissions banking programs and fuel economy bonuses for the manufacturers of methanol compatible vehicles. Before discussing the relative merits of each strategy, however, it is necessary to recognize differences in the types of methanol vehicles which could be introduced in the next ten years. In a July 1987 report entitled, "Cost and Cost Effectiveness of Alternate Fuels In The Transportation Sector" the EPA outlined three basic types of vehicles suitable for methanol use and their current limitations in the transportation sector:

1. Flexible Fuel Vehicle - can operate on any combination of gasoline and methanol. Technology is well demonstrated. Because FFV's could use methanol when available and gasoline at other times, FFV's are particularly attractive as a "transition" vehicle until an adequate methanol fuel distribution system is established.

2. Dedicated M85 Vehicles - can be produced using existing technology . . . could have higher compression ratio than a FFV. These changes increase the fuel efficiency and performance. The general public could not use these before a distribution system is in place, but fleets could. In the shorter run, the FFV is probable more marketable. Once a methanol distribution system is in place, however, a dedicated methanol car offers clear advantages.

3. Dedicated M100 Vehicles - these are designed to take advantage of methanol's properties as a fuel. Design could be optimized (smaller engine, cooling system and much less expensive catalytic converter). This is a fuel efficient car with low emission characteristics. Although most of the research and development for this has been completed, auto manufacturers will require a longer lead time to mass

produce these vehicles. A fuel distribution system is necessary.[20]

Flexible Fuel Vehicles

The controversy surrounding blends led to the introduction of the flexible fuel vehicle. The flexible fuel vehicle is a term used by Ford Motor Company to describe a vehicle capable of running on either methanol, gasoline, or a combination of the two. A flexible fuel vehicle (FFV) is a gasoline burning automobile with the following component changes:[21]

- fuel sensor to monitor alcohol ratio
- fuel system materials
- fuel tank size
- catalyst
- oil formulation

The key to the FFV is an optical sensor, about the size of a deck of playing cards, that monitors the differences in color between methanol and gasoline in a manner similar to devices used by the oil industry in petroleum pipelines.[22] The sensor is connected to the fuel control system, and checks the fuel ratio on a millisecond basis. Because the relationship between gasoline and neat methanol can be measured with precision, it is easy to adjust engine controls between the two. The difference in fuel composition initiates a difference in the output signal from the sensor to various engine control components, such as spark timing, fuel flow rate and EGR (exhaust gas ratio) flow rate.[23]

The "flexibility" of the FFV is the key to methanol's entry into the fuel market. The problem of refueling is eliminated, as the vehicle can be run on gasoline when methanol is not available; yet unlike previously envisioned "dual fuel vehicles," (which had two separate tanks, one for gasoline, one for methanol), flexible fuel vehicles require little design modification. The cost of conversion runs from $50-200 per vehicle, depending on the size of the production run.[24] Ford Motor Company sees FFVs as the logical bridge from straight gasoline to alternate fuels and so has discontinued production of neat alcohol vehicles.[25]

Since the introduction of the first flexible fuel vehicle model in 1985, several evaluation programs have been initiated with the intention of both reducing vehicle emissions to within the EPA guidelines and encouraging the use of methanol as a replacement for gasoline. A Canadian evaluation plan consists of 20 FFV Crown Victorias operated by the Province of Ontario, the federal government of Canada and the New York Research and Development Administration. The California Energy Commission, together with the California Air Resources Board and the South Coast Air Quality Management District operate a total of six FFV Ford Escorts.[26] In July 1987, the President's Task Force on Regulatory Relief announced a plan to order at least 5,000 flexible fuel vehicles to be used in individual states' fleets.[27] A bill proposed by Representative Phil Sharp (D-IN) in the last session of Congress, known as the Alternative Motor Fuels Act of 1987 (HR 3399), would have required the government to purchase as many flexible fuel vehicles "as is practicable" in 1990, mandating increased government purchase and participation in demonstration programs.[28] An amendment to this bill, proposed by Representative Carlos Moorhead (D-CA), would have sought the sale to the public of flexible fuel vehicles, in conjunction with the distribution by ARCO and Chevron of methanol in Southern California.[29]

CAFE

The capability of using either methanol or gasoline makes the flexible fuel vehicle attractive to the consumer once methanol has been introduced into the fuel market, but it still gives little incentive for the automobile manufacturer to produce the FFV on a large scale, nor does it encourage the production and distribution of methanol fuel. For this reason much of the legislation aimed at encouraging the production and use of FFVs uses the incentive of fuel economy benefits.

In response to the oil embargo of the early 1970s, in early 1975 Congress passed the Energy Policy and Conservation Act which established fuel economy standards for automobile manufacturers known as Corporate Average Fuel Economy (CAFE). These standards require that automobile manufacturers produce a total fleet of automobiles whose average fuel economy meets certain mileage standards. Today's CAFE standard is 27.5 mpg, although it has been rising fairly steadily throughout the last decade.[30] Because the standard is based on an average, some cars (i.e. luxury cars) may fall well below the average, as long as the manufacturer also produces a given number of cars with fuel economy levels that are significantly above the average. A company can further amass credits to carry forward or backward for a period of up to three model years.[31] While Japan has exported enough economy cars that they are not constrained in exporting luxury cars for sale in the United States, GM and Ford are having a difficult time meeting CAFE standards.[32]

The potential for methanol introduction into this situation shows an interesting set of incentives. Methanol offers a higher gpm rating than gasoline on a BTU equivalent basis. Further, methanol blends lower the percentage of gasoline that is being used to achieve equivalent mileage. This would increase even further when the fuel used is M85 or M100 in dedicated methanol vehicles. One proposed approach is to give very high mileage ratings to FFVs. Ford and GM have recognized that their inability to meet CAFE standards may have positive effects on the future of methanol-powered vehicles in the U.S., and evidently so have lawmakers.[33] Representative Phil Sharp's Alternative Motor Fuels Act of 1987, in addition to requiring the government to purchase flexible fuel vehicles, would grant CAFE credits to those automakers who build them.[34] At least two other bills currently pending would also provide similar incentives for automakers. Similar legislation also has been proposed in the Senate by Jay Rockefeller (D-WV) whose bill (S.1518) would allow manufacturers, for the first five model years beginning in 1993, the choice of claiming a set increase (1.2 mpg) in the average fuel economy or the average fuel economy of 200,000 methanol (dual-fuel) vehicles, whichever is higher, but not to exceed 1.5 mpg. For the period 1988-1993, the manufacturer could claim a 0.9 mpg increase in economy or the average economy of 200,000 dual fuel passenger cars, whichever is higher but not to exceed 1.1 mpg.[35] Rockefeller's bill has been the most successful of those promoting CAFE credits in gaining Congressional support, and the key to its success seems to be the caps set on the fuel economy to be gained.

Critics of the CAFE incentives for methanol vehicle production say that it gives unfair fuel economy benefits without assurance that methanol will even be used. This argument is especially true with regard to flexible-fuel vehicles. The CAFE law was established in an attempt to limit the use of gasoline--yet giving credits for methanol fuel capability neither ensures the use of methanol fuel (since a FFV can run on gasoline only), nor does it necessarily decrease gasoline use. In fact, it can increase the use of gasoline by granting "empty" credits (mpg) instead of mandating a more efficient automobile.[36]

GM has called methanol the "gasoline of the 21st century;" and at the Sixth International Symposium on Alcohol Fuels in Ottawa, both Ford and GM agreed methanol is the most likely substitute for gasoline[37] Ford has even gone so far as to propose a four-step phase-in program on the part of the automotive industry that could help break the chicken and egg cycle. The basic plan includes:

1. an experimental stage during which a fleet of 5,000 vehicles is introduced (already here)

2. an increase in the experimental fleet to 10,000

3. annual volume increases of up to 30,000, with the production increasing to 50,000 as a regional market is achieved.

4. full commercialization - 250,000-300,000 vehicles a year on an unrestricted basis.[38]

There is a catch, however. Company President Donald Petersen has stated that the company will not proceed beyond the first step until it is certain that the infrastructure to make methanol a major fuel source will be built.[39]

Both the automobile industry and the fuel industry have cited the absence of regulation as a reason for their reluctance to produce both methanol and methanol vehicles.[40] EPA has now issued a final rule that specifies, in 40 CFR Part 86, emissions standards and test procedures for all new motor vehicles operating on methanol fuels. The regulations are effective with the 1990 model year and parallel EPA regulations for petroleum-fueled vehicles for CO, NO_x, particulates, smoke and evaporative hydrocarbons. Similarly, a fuel equivalency factor will be necessary before CAFE credits are practical incentives to the automobile manufacturers.[41] Representatives of the chemical fuel industry also have cited the need for fuel standards as an essential prerequisite for fuel grade methanol. "Fuel grade methanol has yet to be defined. Therefore the industry has not shipped it, and has not sold it."[42]

In addition to setting guidelines for automobile manufacturers and fuel suppliers, the EPA needs to enforce those standards already in existence, such as the NAAQS and CAFE. Rather than waiting to be sued to force non-attainment areas into compliance, the agency should be taking this initiative on its own. One such action, supported by the DOE, that could lessen our consumption of petroleum while promoting the use of methanol would be to increase the average fuel efficiencies to 30 mpg by 1991-92.[43]

California -- A Case Study

While several groups of citizens have acted where the government and its agencies have demurred, the greatest success story in the introduction of methanol as a transportation fuel has come from California, one of the most energy conscious and environmentally cognizant states in the country.

The state of California consumes 12 billion gallons per year of oil, 3% of the world's total.[44] And while it is the fourth largest producer of oil in the U.S., nearly 58% of its supply is imported.[45] Therefore, during the oil crises of the 1970s, the California public became obsessed with obtaining petroleum fuels. The state legislature mandated the investigation of alternative fuels and enlisted the California Energy Commission to find a practical substitute for petroleum fuels.

The California Energy Commission's Alcohol Test Fleet Program was the first large scale testing of alcohol cars. Three different fleets were placed in the service of various state and local agencies, with the cooperation of major automobile manufacturers and fuel suppliers. Fleet One operated as part of the state's Sacramento area motor pool, and consisted of twelve 1980 Ford Pintos -- four converted to run on methanol, four on ethanol, and four control cars (gasoline-fueled). The four methanol vehicles continue to operate in the state fleet, with one vehicle accumulating over 30,000 miles by early 1984.[46] A second fleet included forty 1981 Volkswagen of America Rabbits and pickups, assigned to daily operation in various state fleets. These were the first alcohol-fueled vehicles to be produced on an American assembly line.

Fleet Three consists of forty 1981 Ford Escorts, which are operated as part of the Los Angeles County Fleet.[47] The 1.6 litre engines were converted to methanol using specially modified parts from Ford. Certain components such as fuel tanks, carburetor floats, coating materials and fuel pumps were modified to achieve more durable standards. In affirmation of its confidence in the durability of methanol vehicles, Ford provided a 5-year, 60,000 mile service warranty with this fleet of Escorts. This fleet serves as an advanced test for methanol powered automobiles. Total mileage accumulation neared 1.1 million miles by early 1984, with the lead vehicle, having been driven over 85,000 miles, continuing to show good fuel economy.[48] Fleetwide fuel economy is 26.7 mpg gasoline equivalent, and engine durability is also comparable to gasoline vehicles.

Concurrent with the Energy Commission's alcohol program, the Bank of America in San Francisco initiated its own experimental program. Beginning in February 1980, Ford Motor Company began delivery of a fleet of approximately 200 cars, designed to operate on pure methanol. Fourteen months and one million miles later, Bank of America issued a comprehensive report pronouncing the experiment a total success.[1]

The experimental program of the Bank of America came to several conclusions about pure or "neat" (100%) methanol versus gasoline:

- it is much cheaper in miles per dollar and more efficient in the automobile engine

- it will increase engine life significantly, with approximately the same maintenance costs of a gasoline engine

- it will greatly decrease or practically eliminate exhaust pollutants

- 10% blends, with 95% thermal energy of gasoline, give the same mileage as gasoline (some cars even had better mileage) and four points higher octane rating.

- it can be used neat in present gasoline cars when those parts in contact with methanol are modified at a cost of a few hundred dollars.

- it will greatly reduce the temperature of combustion and hence that of the engine

- it can be used on a smaller, less expensive engine system

- it will allow a new, entirely redesigned methanol car to be built around such an engine, which will be lighter, cost less, and not have a clumsy water radiator and cooling system that is always necessary to eliminate a large amount of waste heat from gasoline combustion.[50]

Bank of America initiated its program in 1979 in response to high oil prices. Since its fleet is centrally fueled it was not difficult to implement a methanol vehicle program, and central fueling of the vehicles allowed BoA to avoid high gasoline prices and long lines, keeping its mail fleet mobile. In addition to the Ford vehicles, BoA purchased 40 General Motors Corp. Chevrolet Citations and Pontiac Phoenix automobiles. To date the bank has put 20 million miles on the methanol vehicles; and retrofitting technology gained by BoA mechanics during the test program enables them to now convert a gasoline engine in up to 16 different car types to methanol use in a matter of hours. BoA found that it saved from

$1,200 to $1,300 on General Motors cars converted by the bank and operated over a 100,000 mile life expectancy.

The Bank has approximately 100 vehicles remaining and has been selling the vehicles, converted back to gasoline as the cars reach their "life expectancy." This practice is standard procedure but the bank doesn't plan to purchase any more while oil prices remain low.

A major aspect of the statewide methanol program has been the cooperation of fuel suppliers. Since the inception of California's methanol program, a chain of about twenty-four methanol stations has been set up by independent marketers under contract.[51] In July 1987 ARCO oil company agreed to begin marketing M85, a blend of 85% methanol and 15% premium unleaded gasoline at 75 stations in California within three years.[52] ARCO further agreed to a CEC request to install by the end of 1988 one pump at each of 25 locations at a cost of $35,000 per installation to dispense M85 supplied by the California Energy Commission.[53]

As elsewhere in the nation, while The Alcohol Fuels Program began as a preventative measure against future petroleum supply disruptions, it has gained recent support as a prescriptive measure. California's experience with alcohol fuels can be seen as a model for what could be our nation's experience as a whole. A member of the Executive Board of California's South Coast Air Quality Management District recently testified before the House Subcommittee on Fossil and Synthetic Fuels:

In 1983, the Federal Air Quality Standard for Ozone was exceeded on 152 days, with maximum ambient levels more than three times above the level set to protect the public health. Federal Air Quality Standards for carbon monoxide and total suspended particulates were also violated as was the average annual standard for nitrogen dioxide. According to the EPA the South Coast Basin is the only area in the country where the NO_x standard is violated. The use of petroleum in our basin--its transport, refining, distribution, and end use--represents over 95% of the reactive organic gases and oxides of nitrogen (NO_x) emissions in the basin. Simply stated, the attainment of federal air quality standard in our basin will require the virtual elimination of petroleum-based fuels.[54]

California's situation is undeniably exceptional. Out of the 76 areas in the nation currently in violation of ozone levels, 11 are in California. Ten of the 81 areas exceeding carbon monoxide standards are also in that state.[55] It is precisely because of the dangerously high levels of air pollution that California has turned to methanol. The South Coast Air Quality Management District (SCAQMD) believes the transition to methanol fuel for all motor vehicles represents the most significant opportunity for air quality progress which exists between now and the end of the century.

Various studies have supported Dr. Berg's belief. A study done by Systems Application Inc. suggests that ozone levels in the South Coast Basin could be reduced to levels near or below the federal standard if methanol fuel were substituted completely for gasoline. Another report, entitled, "Methanol As An Ozone Control Strategy in the Los Angeles Area" used air quality models that projected the substitution of near neat methanol for gasoline in the South Coast Air Basin in forecasts to the year 2000, and drew the following conclusions:

1. Complete substitution of methanol fueled vehicles for gasoline fueled vehicles would lead to a reduction from 14.4% to 20.0% in the peak hourly average concentration of ozone.

2. Peak ozone concentration decreases approximately linearly with the degree of methanol substitution.

3. Photochemical reactivity of methanol (a causative factor in production of smog) is low.

4. The maximum reduction in ozone concentration achievable by elimination of gasoline fueled vehicles emissions is 25%.

5. The results of this study indicate that methanol in motor vehicles is a potential control strategy in future years for areas with severe ozone problems.[56]

These studies, along with the obvious urgency of the need to improve their air quality, have provided the impetus for much legislation aimed at a statewide increase in the use of methanol. By 1984 the CEC had invested almost $10 million in the evaluation, testing and demonstration of methanol as a replacement for petroleum, having leveraged matching investments on the part of private industry in excess of $20 million.[57] In early 1987, Governor George Deukmejian signed a bill providing $5 million for methanol demonstration programs. The CEC has earmarked $2.5 million of this money for the purchase of 3,000-

5,000 flexible fuel vehicles in a joint venture with the General Services Administration.[58]

Several bills involving methanol are currently before the California legislature, including the following:

- a bill that would empower air quality management districts in non-attainment areas to require stationary sources of pollutants to switch permanently to clean fuels. This bill would provide funding for the study of the potential of clean fuels for the long term air pollution reduction and would allow utilities to pass the added cost of switching fuels to rate payers

- a bill that would use some of California's highway funds for the purchase of methanol buses and establish a SCAQMD clean fuels office to develop a five year program to promote methanol

- a measure that would stipulate that beginning in 1990, no less than 10% of light duty vehicles be limited to emissions of 0.2g HC or be capable of running on methanol. This portion would increase to 25% by 2000.

- a measure that would require fleet operators in the state to have a specified share of FFVs (flexible fuel vehicles) in fleets and fuel them with methanol or offer the option of obtaining air emissions offsets by buying FFVs. (This measure could account for 10,000 to 20,000 FFVs/yr. out of total California sales of about 500,000 vehicles/yr. FFV would add about $50-200 to cost of the car.[59]

On January 8, 1988 the South Coast Air Quality Management District approved an extensive strategy intended to replace up to 400,000 diesel and gasoline powered fleet vehicles with those that run on cleaner burning fuels or electricity beginning in 1993. All public and private fleets with fifteen or more vehicles, including public transit bus systems and rental car agencies, are included in the plan. Furthermore, in an unprecedented move, all new (1988 and onwards) electrical turbines, internal combustion engines and cogeneration units must use methanol as a backup fuel instead of diesel. The $30.4 million cost of the clean fuels program will be raised from increased permit fees charged industrial polluters, a surcharge on motor vehicle registration fees, matching funds from private industry for demonstration projects and financial assistance from government agencies.[60]

The clean fuels edict marks the first legislation aimed at fuel combustion itself rather than emissions levels, and many believe the approval of the strategy will increase the possibility of a shift from gasoline and diesel to clean fuels for 40% of the region's passenger vehicles and 70% of the trucks. The Air Quality Management District also plans to present a near-term schedule to phase out the use of diesel fuel in all stationary boilers, turbines, and other equipment.

The Department of Energy Scenarios
From Theory to Practice

The success of California's experiment with methanol proves that with a will, the chicken and egg cycle can be broken, and a methanol program can succeed. Once initiated, all that remains are the logistics of implementation. Worldwide there are 450 million vehicles to turn over in the conversion process to alternate fuels.[61] While at first glance this may seem an impossible task, two facts bode well for a potentially smooth transition. The first is that the U.S. produces one fourth of the world's vehicles, more than any other country, and is unmatched in the pace at which it adds to its fleet.[62] Furthermore, it leads the world in the volume and value of the transportation fuels it consumes, which means that trends in technology here have a great effect on world markets. Consequently, the introduction of methanol technology into our automobile fleet could spread fairly quickly. A second fact that is encouraging for the implementation of methanol technology is that catalytic converters, which became the standard along with the phaseout of lead, were added to 75% of all U.S. vehicles in only one year. Furthermore, the total cost of implementation for this new environmental equipment shows surprisingly little variation in whether the turnover takes place very gradually or on an accelerated basis.[63]

Costs

The lifecycle cost of methanol vehicles versus gasoline vehicles is largely determined by the price of the fuel and the comparative cost of each vehicle. Any car can be converted to run on methanol by adjusting the spark timing and fuel-air mixtures and replacing the plastic and rubber parts that the fuel can dissolve. The cost of conversion is about $1,000.[64] The cost of a mass produced methanol vehicle can be as much as $750 more than a gasoline vehicle if the production run is small, but large production runs would produce methanol vehicles comparable in price to gasoline vehicles.[65]

Distribution

Thousands of service stations across the nation are currently equipped to handle blends, and there are about 30 methanol dispensing systems presently in use, with some dating back to 1981 and earlier; so the distribution of methanol at retail outlets should pose no practical problems.[66] It has been estimated that about 93,000 service stations account for 77% of all retail fuel sales in the U.S.[67] These stations could be equipped with methanol fueling units for $28,000 each (the figure used by ARCO in its agreement with the California Energy Commission). However, because half of this cost is associated with the installation of the fuel tank, this cost could be substantially reduced by using existing tanks. Currently about 90% of the retail tanks in service are made of carbon steel, which is compatible with methanol.[68] Recently many tanks are being replaced because of concerns over environmental effects, mainly leakage; and these are being replaced with fiberglass, also suitable for methanol. This recent replacement of tanks would further facilitate the introduction of methanol.

Two Scenarios

One Million Barrels Per Day

In January 1988 the Department of Energy issued a report entitled, "Assessments of the Costs and Benefits of Alternate Fuels in the Transportation Sector," which reported, "DOE has evidence that to affect world oil prices significantly by the turn of the century, an increase in transportation fuel flexibility would have to be capable of displacing at least a million barrels of oil per day around the year 2000."[69] The Department of Energy then examined the feasibility of displacing 1 MMB/D by the substitution of methanol. The report concluded that to displace 1 MMB/D oil, 30 million vehicles would have to start using methanol (or some other fuel). This would require 30 billion gallons a year of methanol, an amount that could be produced if an average of ten methanol plants were to come on line every year for the next ten years. Between 10,000 and 30,000 service stations would be required to open each year during this ten-year period, to provide a total of 150,000 methanol dispensing units. Simultaneously, 1.5 to 8 million methanol compatible vehicles would enter the fleet after a 3 to 5-year period of redesign and retrofitting vehicles.[70]

The cost of implementing such a program, assuming no drastic change in fuel price, would be approximately $30 billion for the production facilities ($20-25 billion) and service stations ($5 billion). Effects on the existing infrastructure would likely be: the slowing of an oil price increase (since alternate fuels would be coming into market as they become attractive due to the rise in oil prices), a potential drop in gasoline prices, (owing to competition for customers in the face of excess capacities) and the fall of refiner profits as methanol assumed a greater share of the fuel market.[71]

As is usually the case with an innovation, it is difficult to assign a monetary value to the benefits of such a level of oil displacement. The DOE report comments, "Some of the gains envisioned involve possible economic and security advantages that need not be proportional at all to expenditures by vehicle purchasers and users who had financed the changeover. Instead the benefits (which could easily be worth billions of dollars annually in all) would come to the nation at large."[72]

The Two Million Barrels Per Day Scenario

"If alternative fuel use did cut U.S. transportation demand for oil by as much as 2 MMB/D the rule of thumb . . . implies that world oil prices might be reduced by about $4.00/barrel from what they otherwise would be . . . . This could cut total national outlays by tens of billions of dollars annually . . . ."[73] The displacement of one million barrels of oil daily is much more likely in the short term, but once accomplished, we have the feedstock for a larger displacement, and the transition to 2 MMB/D methanol use will be smoother than the initial displacement because the infrastructure will already be in place. At a November meeting of the DOE Energy Research Advisory Board, the capability of meeting the potential for a 2 MMB/D oil displacement was examined closely. While it has not been promoted as fervently as the 1 MMB/D scenario, the practical replacement of 2 MMB/D of oil is by no means unrealistic.

Assuming the displacement of 2 MMB/D of oil in the year 2000 in favor of M85 (one gallon M85 displacing 0.4 gallons gasoline) used in dedicated and flexible fuel vehicles, the Energy Research Advisory Board found total infrastructure costs (not including production facilities) to range from $6.1 to $11.2 billion dollars (1986 dollars) for the introduction of methanol into the fuel market over the nine-year period from 1991-2000. The breakdown was as follows:

$2-10 million per terminal
$5-35 thousand per retail outlet
$50-200 per car

Over the nine-year period from 1991 to 2000, a total of 1.5 to 17.9 million new M85 vehicles per year would come into the automobile market for a total of 91 million vehicles (with 75% access to M85). These would be fueled by retail fuel outlets coming on line at a rate of 12,000 to 58,000/yr for a total of 344,000 retail outlets, each providing around 200,000 gallons/yr. M85 for a total fuel capacity of 73 billion gallons/yr.

The potential benefits of such a displacement depend largely on the relative prices of gas and methanol, but fall into two general categories: consumer savings from lower fuel costs, and national savings from the reduced price of imported oil. Furthermore, the hidden costs associated with gasoline will be reduced as is our dependence on it, shaving off a substantial amount from our trade deficit.

Chapter Five: Summary

Methanol can offer both environmental benefits and a buffer against oil supply disruptions; but until it is widely available, it can do little good. Fuel suppliers do not wish to sell it until they are guaranteed a market, while automobile manufacturers are reluctant to mass produce methanol vehicles before a fuel supply is in place. This situation, common to almost any innovation, is known as "the chicken and egg dilemma."

Demonstration programs are one way in which methanol's use can be promoted. The Department of Energy has sponsored fleets in Berkeley, California; Argonne, Illinois; and Oak Ridge, Tennessee. As of June 1987, 29 vehicles were operating in two DOE fleets with little difficulty, and reductions were found in both carbon monoxide and nitrogen oxide emissions. Yet these demonstration programs, which have been possible in part because government fleets can be centrally fueled, have done little to promote the adoption of methanol within the commercial sector.

Attempts at introducing methanol to the commercial market are targeted at either the fuel industry or the automobile industry. Because the current administration opposes market intervention, fuel industry incentives have been linked to the Clean Air Act, and mandate alcohol blends in non-attainment areas. Supporters of this strategy argue it is a slower, less disruptive method of introducing alcohol to the market, while opponents say it does little to encourage the use of 100% alcohol as a fuel in itself.

Incentives for the production of methanol-burning automobiles focus mainly on the flexible fuel vehicle, capable of running on any combination of gasoline or methanol, and include fuel economy credits to the manufacturers who build them. While proponents of this plan point out the advantage of creating an automotive infrastructure capable of running on methanol, critics say it will offer credits to manufacturers without the guarantee that the vehicles will use methanol.

While the arguments against and in behalf of various aspects of "chicken and egg" solutions seem as endless as the dilemma itself, the state of California has taken its own initiative in promoting methanol with surprising success. With nearly 58% of its oil supply imported, and some of the worst air quality in the nation, California has good reason to search for an alternative fuel. Since 1979, over 200 vehicles have operated on pure methanol in the Golden State and many more on M85. Within the next three years M85 will be available at 75 gas stations in California, and recently a plan was approved to replace up to 400,000 diesel and gasoline vehicles with ones burning clean fuels.

The Department of Energy has projected two scenarios in which methanol could replace a portion of our fuel needs. The lower scenario, projecting a one million barrel a day substitution of methanol by the year 2000, would amount to $30 billion in infrastructure costs including about $25 billion for methanol production, to fuel 30 million vehicles with methanol, a feat whose benefits could be worth billions of dollars annually according to the DOE. A two million barrel per day oil displacement, while less likely, would cost between $6.1 and $11.2 billion for the vehicles and the fuel distribution infrastructure, and could effect a decrease in oil prices of about $4.00/barrel. Either of these strategies, while expensive at first glance, would lower fuel costs to consumers, and reduce the trade deficit of the nation as a whole.

Chapter Five: Vital Statistics

Number of methanol vehicles (M85) operating in U.S. Federal fleets as of June 1987: 29.

Mileage accumulation of Fleet Three, Los Angeles County Ford Escort (M85) vehicle fleet, by 1984: 1.1 million miles

Bank of America fleet of pure methanol vehicles:

Number	100
Mileage accumulation	10 million miles

Cost of conversion on a per vehicle basis from gasoline powered vehicle to Flexible Fuel Vehicle (varies according to size of projection run): $50-200

Number of FFVs operated by the California Energy Commission: 6

Number of gas stations in Southern California that will market M85 by 1990: 75

Cost of conversion of a single gasoline powered automobile to run on methanol (M85): $1,000

Cost per vehicle of conversion for small production run: $750

Cost of methanol vehicle (M85) based on large production run: comparable to gasoline powered vehicles

Cost per retail outlet of equipping fueling stations with methanol tanks: $28,000

To replace 1 MMB/D of oil:

Number of vehicles that need to start using methanol: 30 million

Number of dispensing units: 150,000

Cost of such a program: $30 billion

To replace 2MMB/D of oil:

> Number of vehicles that need to start using methanol: 91 million

> Number of retail outlets supplying methanol: 344,000

> Number of gallons of methanol needed: 73 billion gallons/yr.

> Infrastructure cost (less fuel production): $6.1 - $11.2 billion

> Possible effect on oil prices: $4.00/barrel reduction

Chapter Five: End Notes

[1]E. Eugene Ecklund and Ralph N. McGill, "Introducing Methanol-Fueled Vehicles into Government Fleet Operations," conference paper submitted to the VII International Symposium on Alcohol Fuels, Paris, France, October 20-23, 1986.

[2]*Ibid.*

[3]United States, General Accounting Office, "Alternate Fuels: Information on DOE's Methanol Vehicle Demonstration Program" 1987, p. 9.

[4]Ecklund and McGill, *op. cit.* Conversions were made by the Bank of America and included: fuel line replacement (Carburetor); electroless nickel plating (carburetor); enlarging fuel metering jets; replacing head gaskets; a larger fuel tank; no change in compression ratio.

[5]US GAO, *op. cit.,* p. 10.

[6]Ecklund and McGill, *op. cit.*

[7]*Ibid.*

[8]This is believed to be the result of the wrong types of rings and spark plugs initially installed.

[9]US GAO, *op. cit.*

[10]*Ibid.,* p. 7.

[11]US GAO, October 1984, p. 12.

[12]Anderson, "Lead Cut Gives Alcohols Crack at Gasoline Blend Market," *Chemical and Engineering News*, April 8, 1985, p. 18. Cosolvents are ethanol, propanols, butanols or any mixture.

[13]U.S. Department of Energy, "Assessment of Costs and Benefits of Flexible and Alternative Fuel Use in the U.S. Transportation Sector," January 1988, p. B-5.

[14]Office of the Press Secretary, The White House, "Fact Sheet: President Bush's Clean Air Plan," June 12, 1989, 14 pp.

[15]*Ibid.,* pp. 8-9.

[16]Larsen, R.P. and D.J. Santini, "Rationale for Converting the U.S. Transportation Sector to Methanol Fuel," May 1986, n. p.

[17]*Ibid.*

[18]Although the original request for a waiver to use methanol as a blend in gasoline was restricted by the EPA because of this volatility concern, reconsideration has led the EPA to the view that the blend will not increase volatility beyond current fuel indexes. "EPA Clears Way for More Alcohol Fuels," (*Chemical & Engineering News,* November 3, 1986, p. 23.)

[19]Information Resources, Inc., "Understanding the Challenges and Future of Fuel Alcohol in the United States," prepared for the U.S. DOE Office of Alcohol Fuels, September 1988, p. 165.

[20]U.S. Environmental Protection Agency, "Air Quality Benefits of Alternative Fuels," June 1987, p. 4

[21]Department of Energy, Notes from the Energy Research Advisory Board Quarterly Meeting, November 1987, p. 8.

[22]"Flexible Fuel Vehicle Could Help Popularize Benefits of Methanol," *Research and Development,* August 1986, p. 33.

[23]Ford Motor Company response to Specific Questions from Rep. Sharp regarding H.R. 3355, Material Submitted for the House Subcommittee on Fossil and Synthetic Fuels, November 20, 1985.

[24]Department of Energy, Energy Research Advisory Board, November 1987, p. 8.

[25]"Sharp Methanol Bill Amendment Seeks Government FFV Sales to Public," *Alcohol Week,* November 23, 1987, p. 5.

[26]*Research and Development, op. cit.,* p. 33.

[27]Taylor, Robert E. "U.S. Plans to Buy 5,000 Vehicles that Use Methanol, *Wall Street Journal,* July 15, 1987, p. 2.

[28]"House Approves Sharp Methanol Bill By Big Margin Despite Veto Threat, *Alcohol Week,* December 21, 1987, p. 2.

[29]*Ibid.*

[30]Scheiba, Shirley Hobbs, "Unwise at Any Speed: Those Absurd Gasoline Mileage Standards Should be Scrapped," *Barrons,* August 4, 1986, p. 9.

[31]*Ibid.*

[32]*Ibid.*

[33]Sundstrom, Geoff, "Ford, GM Report on Methanol Cars," *Automotive News,* July 29, 1986, p. 31.

[34]*Alcohol Week,* December 21, 1987.

[35]"Rockefeller Calls on CAFE Credit Caps 'Key' to Passing Methanol Bill in Senate," *Alcohol Week,* December 14, 1987, p. 11.

[36]Kahn, Helen "Methanol Trendy Among Lawmakers," *Automotive News,* December 2, 1985, p. 8.

[37]Anderson, Earl, "Methanol Touted as the Best Alternate Fuel for Gasoline, *Chemical & Engineering News,* June 11, 1984, p. 14.

[38]*Ibid.* p. 15.

[39]*Ibid.*

[40]"Lack of Methanol Vehicle Standards Slows Production for Consumers," *Alcohol Week,* November 30, 1987, p.4.

[41]Kahn, Helen, "Methanol Becoming a Pet in Congress, But Problems Loom," *Automotive News,* April 9, 1984, p. 2.

[42]Dixon, R.H. "Methanol: A U.S. and Global Picture," Oxygenated Fuel Conference, Arlington, VA, November 18, 1982.

[43]Information Resources, Inc., *op. cit.,* p. 160.

[44]Williams, Bob, and Michael Obel, "Air Quality Concerns Buoy Hopes for U.S. Makers of Alcohol Fuels," *Oil and Gas Journal,* February 9, 1987, p. 13.

[45]Jackson, M.D., Powers, C.A., and Fond, D.W. "Methanol Fueled Transit Bus Demonstration," Paper no 83-DGP-2, American Society of Mechanical Engineers.

[46]Smith, Kenneth D., Dan W. Wong, Don S. Kandoleon, Cindy A. Sullivan, California Energy Commission, "The California State Methanol Program-Creating Market," and "Methanol as Ozone Control Strategy in the Los Angeles Area," California Energy Commission, Sacramento, California, presented at the VI International Symposium on Alcohol Fuels Technology, May 21-25, 1984, Ottawa, Canada.

[47]California Energy Commission, brochure, "Methanol . . . The Fuel of Tomorrow . . . Here Today," p. 4.

[48]Smith, Wong, et al., *op. cit.*

[49]"Methanol: America's Answer to Motor Fuel Independence," A presentation by R. Jack Alexander to governing boards of Automotive Warehouse Distributors Association and Motor Equipment Manufacturers Association, Carlsbad, California, June 1982.

[50]Othmer, Donald F. "Methanol Fuel for Automobiles," *Chemical Engineering Progress,* October 1985, p. 17.

[51]Emond, Mark, "Is Methanol Marketing Niche Shaping Up in the West?," National Petroleum News, August 1987, pp. 18-19.

[52]Shaner, J. Richard, "Methanol Motor Fuel Looking Good Again," *National Petroleum News,* July 1987, p. 14.

[53]"Arco to sell Methanol Fuel in California," *Chemical and Engineering News,* June 1, 1987, p. 5.

[54]Testimony of Dr. Larry Berg, Member of the Executive Board, South Coast Air Quality Management District, before the House Subcommittee on Fossil and Synthetic Fuels and Subcommittee on Energy Conservation and Power, April 4, 1984.

[55]U.S. EPA, *op. cit.,* p. 3.

[56]Wilson, K.W., and McCormack, M.C., California Energy Commission, "Methanol as an Ozone Control Strategy in the Los Angeles Area," presented at the VI International Symposium on Alcohol Fuels, Ottawa, 21-25 May, 1984, Attachment III.

[57]Ford, General Motors Company, Celanese Chemical Corporation, Volkswagen, Conoco and DuPont were those corporations primarily involved. Letter from Charles Inbrecht, Chairman of the California Energy Commission, to Rep. Phillip Sharp, Chairman of House Subcommittee on Fossil and Synthetic Fuels, April 3, 1984.
[58]Williams, Bob, Michael Obel, "Air Quality Concerns Buoy Hopes for U.S. Makers of Alcohol Fuels," *Oil and Gas Journal,* February 9, 1987, p. 15.
[59]*Ibid.,* p. 14.
[60]Stamner, Larry B. "AQMD OKs Plan to Cut Use of Diesel Gasoline," *Los Angeles Times,* 1-9-88. p. I-1.
[61]Anderson, E. *Chemical and Engineering News,* June 11, 1984, p. 16.
[62]Department of Energy, *op. cit.,* January 1988, p. 32.
[63]*Ibid.,* p. 37.
[64]Taylor, Robert E., "Methanol Advocates Spark Interest in the Fuel But Critics Say Potential Benefits Will Vaporize," *Wall Street Journal,* September 4, 1987, p. 38.
[65]EPA, "Cost and Cost Effectiveness of Alternate Fuels," July 14, 1987, p. 2.
[66]Mills & Ecklund, *Annual Review of Energy,* 1987, p. 61.
[67]*Ibid.,* p. 62.
[68]*Ibid.*
[69]DOE/PE, *op. cit.,* p. 4.
[70]*Ibid.,* pp. 35-37.
[71]*Ibid.,* p. 38.
[72]*Ibid.,* p. 34.
[73]*Ibid.,* p. 27.

CHAPTER SIX

COMMERCIAL PRODUCTION OF METHANOL: USING ALL OUR POTENTIAL

Based on our knowledge today, it is probable that in 50 years we'll be driving on methanol, more probably than any other fuel.

-Ernst Fiala

Once methanol is introduced in volume into the transportation sector, it can offer several benefits to our energy supply beyond being an obvious alternative to gasoline. Because of its relatively easy synthesis from carbon, hydrogen and oxygen, it can be produced from a broad variety of feedstocks, giving the U.S. increased resource flexibility in the event of a supply disruption and oil price increase. Furthermore, as the cost of conventional feedstocks increases over time, more environmentally desirable alternatives should become economically competitive. This chapter will look at the history of methanol production, examine current production technologies and capabilities in the U.S. and the world, assess the ability to scale-up U.S. production to meet the demands from a major share of the transportation sector, and look ahead to see the methanol production technologies of the future.

History of Methanol Production

For over two centuries, until shortly after World War I, methanol was produced by heating wood in the absence of air (called destructive distillation), a process that resulted in its common name, wood alcohol. Early processes distilled wood alcohol from a liquid that was produced during the manufacture of charcoal. Using that process, one ton of wood could produce three to six gallons of methanol.[1] In 1905, the French chemist Paul Sabatier suggested that methanol might be produced synthetically, by combining carbon monoxide and hydrogen. His process was adapted by the Badische Company, which set up the first synthetic commercial plant at Leunawerke, Germany in 1923.[2] In February 1924 they began exporting synthetic methanol to the U.S. at a cost of only two-thirds the cost of wood alcohol, thus signaling the end of a major wood-alcohol industry, and prompting the production of synthetic methanol in the United States.

The Commercial Solvents Corporation began domestic production of methanol in the U.S. in 1926, using hydrogen and carbon dioxide formed during the fermentation of corn; but soon the company was using coal as their source for the synthesis gas mixture.[3] The Haber process, as it was called, involved the combustion of coal to produce a gas called synthesis gas that was then converted to methanol by reacting it with a chromium-oxide and zinc-oxide catalyst under very high pressure and high temperature.[4]

Synthesis gas is a mixture of hydrogen, carbon monoxide and carbon dioxide that can be produced by the partial oxidation of any fuel containing carbon with oxygen or water. Synthesis gas from coal has a less than optimal hydrogen content and thus has to be treated in a manufacturing plant. Natural gas is hydrogen rich; therefore, in the production of methanol from natural gas, excess hydrogen is usually used to make ammonia in an adjacent plant. This characteristic, along with the expansion of the petroleum industry, led to the replacement of coal with natural gas as a feedstock for methanol production.[5] Today, over 90% of the world's methanol production capacity is designed for natural gas as a feedstock because of its higher relative hydrogen content and lower capital and operating costs.[6]

Methanol Production Process

The production of methanol from natural gas is a two-step process. The first step, "reforming," involves heating sulfur-free natural gas, CH_4, with steam over a catalyst to produce synthesis gas. The equation looks like this:[7]

$$CH_4 + H_2O \ \text{-----} > \ CO_3OH + H_2 + CO_2$$

The reformer represents the largest and most expensive component in a gas to methanol plant. The heat given off from this reaction can be captured and used elsewhere in the plant as an energy source.

Once the correct composition of synthesis gas is obtained, the second step, the "synthesis" of the fuel is achieved in the presence of a chromium-oxide and zinc-oxide catalyst at a temperature of about $660^\circ F$ and a pressure of about 300 atmospheres.

$$CO + 3H_2 \ \text{------} > \ CH_3OH + H_2$$

Since the optimum carbon-hydrogen ratio is 1:2 and natural gas is 1:4 (CH_4), carbon dioxide is usually added to take care of the surplus hydrogen, causing a reaction that produces a liquid containing methanol and water.[8]

$$CO_2 + 3H_2 \quad ------> \quad CH_3OH + H_2O$$

The raw methanol produced contains about 15% water by weight and traces of other impurities. Therefore one final step, distillation, is necessary for most uses of the liquid.[9]

The earliest processes developed for the synthesis of methanol operated at relatively high temperatures (300°-400^oC) and pressures (275-360 atm). In 1966 the single most important improvement in the production of methanol occurred when Imperial Chemical Industries (ICI) of the U.K. developed a low pressure synthesizing process using a copper-based catalyst.[10] This development permitted a much lower pressure (50 atmospheres) and temperature (500^oF) for the reaction of the synthesis gas and the catalyst. Capital and operating costs were reduced as a consequence of the lower temperatures and pressures needed, and the need to add CO_2 to balance the C:H ratio was eliminated.[11]

Today the most common low pressure processes are the ICI method and West Germany's Lurgi process, which reduces methane consumption 8% and synthesis gas 20% per ton of methanol produced by combining the reforming and synthesis processes described above.[12] Virtually all commercial plants today use a zinc-copper catalyst in one of these two processes.[13]

Methanol and Today's Market

While methanol's use as an automotive fuel has been successfully demonstrated, it is generally known for its use as a chemical. Currently in the U.S., 90% of methanol consumption is for conventional chemical production, such as formaldehyde, acetic acid, acetic anhydride and various chemical solvents.[14] In the past thirty years the capacity of the domestic methanol industry has expanded tremendously, from 300 million gallons in 1960 to 1.1 billion gallons in 1979 to 1.9 billion in 1985.[15] Yet even with the rapid growth of the chemical industry, its capacity would need to be increased significantly to supply a meaningful portion of the transportation sector's needs. Total 1986 U.S. methanol production for all uses was about 1.2 billion gallons, a quantity equivalent in energy to about 16 million barrels of crude oil, or about 0.5% of the 1986 energy consumption of the transportation sector. Some 280 million gallons per year of methanol production are currently used in the transportation sector, an amount equal to 0.1% of that sector's energy demand. Of the methanol used in the transportation sector, about 95% is used as a feedstock for the gasoline additive, MTBE, which is then blended with unleaded gasoline to increase its oxygen content.[16]

Current feedstock capacity to increase methanol production significantly is not a problem. However, instability on the production side has raised some concerns. U.S. capacity has gone from a record high of 1.77 billion gallons of methanol production in 1982 to 1.08 billion gallons in 1987, a reduction of almost 40%.[17] This occurred as a worldwide glut, around the end of 1985, led to price decreases. The price of methanol plunged from around 38 cents a gallon in the first quarter of 1986 to 26 cents a gallon by year end.[18] The excess supply and falling prices led to several plant shutdowns around the U.S. and worldwide. In 1987, the demand for methanol increased (reflecting the rapidly growing market for MTBE as an octane enhancer), tightening supplies and sending prices up 50% or more in the U.S. and more than 40% in Europe.[19]

These wild swings in the supply/demand balance -- normalcy in 1982, over supply in 1985-86, and a recent tightening and acute undersupply in the U.S. -- has significance beyond the immediate price surges. One of these is imports. Imports have risen from a level of 62 million gallons in 1982 to an estimated level of over 460 million gallons in 1988, an increase of over 700%.[20] This figure represents over one-third of our annual methanol consumption. While the majority of our imports come from friendly nations, (Canada, Trinidad and Chile), our reliance on foreign sources today can have only negative implications for our energy independence tomorrow.

The primary constraint on increased production of methanol is plant capacity. The U.S. currently has a total of 17 methanol production plants, although not all open or operating at full production capacity.[21] To supply only 1 MMB/D of methanol to offset U.S. transportation fuel consumption, over 100 new plants would be needed, each producing an average of 250 million gallons a year.[22] Yet several considerations face manufacturers of methanol. The cost of "demothballing" or reopening a plant that has been shutdown is considerable. In 1986 DuPont reactivated its Beaumont, Texas plant, expecting it to cost between 8 and 10 million dollars. While it is difficult to assess with certainty the final expenditures, by some estimates it cost the company three times that.[23] Furthermore many existing methanol manufacturers are reluctant to turn production away from the certain chemical market toward a new and uncertain fuel market.[24]

While manufacturers may have reservations about increased production of methanol for fuel use, the surge in prices of chemical methanol will undoubtedly spur an increase in production capacity, and this will help to create an infrastructure for fuel supply. Furthermore, the demand for fuel methanol is expected to double by 1990, accounting for 20% of total U.S. production by that time according to economists studying fuel alcohol markets.[25]

Production Capacity

Natural Gas

Currently natural gas supplies are abundant, and prices are low. But as supplies shrink (new discoveries are not replacing consumption-- estimates of worldwide supply were revised downward 5.8% in March 1987)[26] or the price of oil rises, natural gas prices will begin to rise as well. While a price increase will naturally take its toll on the currently low price of gas-based methanol, there is enormous potential for the production of methanol from another source. Natural gas transportation requires construction of expensive pipelines for overland movement and even more costly LNG facilities and equipment for sea movement. There are large quantities of otherwise wasted, and very low-cost gas supplies-- remote natural gas--simply due to its inaccessibility.

Remote natural gas is gas that is far from conventional distribution systems. Often this is offshore, or in areas so far from the market for fuel that it is uneconomical to transport it. One opportunity to use such gas is located near oil drilling operations, where it can be used to repressurize wells or provide the necessary energy for secondary recovery operations. If it must be transported over water to find a market, however, the economics become much less favorable than the cost of oil transportation.

Remote natural gas, if it is marketable, is usually liquefied at the source; and the resultant liquefied natural gas (LNG) is transported in specially designed LNG tankers to distribution ports equipped with LNG regasification facilities, which prepare the gas for entry into a pipeline system for distribution to users. While this method of liquefying the gas and then regasifying the liquid can deliver almost 90% of the energy extracted from a gas well, the system involves very large capital requirements, namely expensive plants at each sea terminal and large ships that are specially designed and constructed for the task. There are substantial hazards in all stages of the LNG process.[27]

In other cases, the inadequacy of local distribution systems to handle such large quantities of remote gas, coupled with an infrastructure insufficient to support the large plants that would use the gas, make it uneconomical for sale, both on local and distant markets. In these circumstances, gas historically has been flared (burned) as a safety precaution or reinjected into the ground as a means of repressurizing the wells and facilitating the recovery of the remaining oil.

An estimated 20 billion cubic feet of natural gas is now flared daily worldwide, an amount equal to over three million barrels of oil a day.[28] The idea of a floating methanol plant at offshore drilling sites has been suggested as a way to make use of gas that otherwise might be flared--as a means of delivering methanol to distant markets more cheaply than LNG. The concept of a barge-mounted methanol plant (BMMP) is not unlike an onshore facility. The barge itself could be made of steel or reinforced concrete, compartmentalized for buoyancy and/or storage, and the methanol plant could be built on or through the deck of one end of a barge. The crew's quarters, a helipad and the loading point for the natural gas would be at the other end in case of a fire. A storage facility, capable of holding between 20 and 30 days of methanol, would be located in the center of the barge. The methanol production facility design would be essentially the same as that of an onshore facility, with necessary modifications for the stresses of the sea (wind, rough seas, etc.). At twenty to thirty-day intervals, when the storage facilities were full, another barge or tanker would off-load the methanol and bring it to port in the same way oil is transported.

The advantages of methanol production over the liquefaction and transportation of LNG can be substantial. Lower capital investment, shorter development time and increased flexibility due to the facility of relocation are the primary benefits of this approach. Furthermore, LNG facilities need large economies of scale to be efficient and cost competitive, while methanol plants can be much smaller and can make smaller gas reserves economic to produce.

The dilemma facing the transport of remote natural gas is best exemplified in the case of Alaska's North Slope, where the costs of reinjection have grown sharply from the 15 cents/mcf of a decade ago.[29] To bring the 31 trillion cubic feet of natural gas to market would require a 4,800 mile pipeline with a price tag of up to $25 billion (1981 dollars).[30] The conversion of this gas to methanol would provide several advantages:

1. Conversion to methanol would use all 2 billion cubic feet/day, while a gas pipeline would be unable to accommodate the estimated 100,000 bbl/day LNG. The remainder would still have to be reinjected.

2. Methanol could be piped through the unused portion of the already existing Alyeska pipeline, whereas LNG requires construction of a separate pipeline specifically for gas.

3. A significant portion of the estimated cost of a system to bring the gas to the lower-48 states would be that of a facility required to remove from the gas CO_2 and higher hydrocarbons which would otherwise condense in the pipeline. By building a methanol facility near the well, not only would this step be eliminated, but the amount of the product methanol would be significantly greater than the natural gas that had been removed of its CO_2 and higher hydrocarbons.[131]

While the conversion of gas to methanol would suit Alaska's situation specifically, the benefits of converting gas to methanol apply to other northern areas nearby as well. Arctic Canadian gas could also be brought south in this way, a potentially significant contribution considering that Canada contains 100 Tcf of proved gas reserves.[32] The energy used in bringing methanol to market would be less than half that used by piping the natural gas.[33]

Finally, methanol's lower capital cost offers greater flexibility of scale in its utilization of natural gas values. To be economically viable a source of at least 250MM cubic feet per day (cfd) of gas is required for the transport of LNG. A BMMP can handle considerably lower gas production rates, with the largest BMMP designed so far requiring close to 100 MMcfd for a completely autonomous unit producing 3000 mtd (approximately 1.2. million gallons/day) of methanol.[34]

The United States House of Representatives Subcommittee on Fossil and Synthetic Fuels supports remote natural gas as the most economical feedstock for the production of methanol. A minimum size plant, producing approximately 2 million gallons a day is estimated to cost between $300-500 million dollars (1988 dollars) in construction costs. Methanol produced at such a facility is projected to cost between 50 and 72 cents a gallon.[35] More recent studies project costs could be 30 cents per gallon or lower.

Coal

Remote natural gas, including Alaskan North Slope gas, is currently considered the best feedstock for methanol because it is otherwise valueless. So, domestically, it is the cheapest feedstock. However, as discussed in Chapter 1, nations other than the U.S. hold the majority of the world's gas reserves and will always be able to provide lower cost sources of natural gas. For this reason coal is seen as the likely candidate for methanol in the long run. Cheap, abundant and significantly cleaner in application when used to produce methanol, coal used for methanol production may be the key to U.S. energy independence.

The primary advantage of a coal-based energy economy as discussed in Chapter 2, is the plentiful U.S. supply and hence low cost. But for environmental reasons (its high sulfur content) and practical considerations (its bulky form and restricted applicability--primarily electricity generation) the U.S. has been unable to capitalize on the fuel source with which it is most abundantly endowed. The conversion of coal to methanol lessens both the problems of form and environmental liability, and could extend our fuel supply hundreds of years.

The United States has twenty times more energy stored in coal than the Middle East has in oil. Estimated U.S. coal reserves are equivalent to 11,751 billion barrels of oil.[36] While the U.S. would need to increase by nearly a factor of 100 its current 4-8 million gallons per day of methanol production capacity to meet current daily consumption of gasoline (approximately 273 million gallons equivalent methanol using a 2:1 energy ratio), we have the necessary feedstock.[37]

The first methanol derived from synthesis gas used coal as a feedstock. However, the rise of the petroleum industry and natural gas' higher hydrogen content led to its gradual replacement with this current feedstock. However, as natural gas supplies shrink, and its price increases, interest in the manufacture of methanol from coal has picked up again.

Beginning in the 1920s methanol was made by reacting coal with steam and oxygen to produce a synthesis gas of hydrogen and carbon monoxide.[38] This traditionally produced a syngas with a $H_2:CO$ ratio

$$C + H_2O \ \text{-----}> \ CO + H_2$$

of 0.7:1; and since this falls short of the optimum 2:1 ratio, hydrogen was traditionally added to achieve the ideal ratio. Recently research efforts have yielded a short-cut to achieving the proper $H_2:CO$ ratio by reacting carbon monoxide and water, to give

$$CO + H_2O \ \text{-----}> \ H_2 + CO$$

Less capital is needed, and more methanol is produced at lower pressures.[39]

In order to produce methanol from coal, the coal must first be gasified (see Figure 6-1). At that stage a molecular sieve permits the

Figure 6-1

Coal Gasification Processes

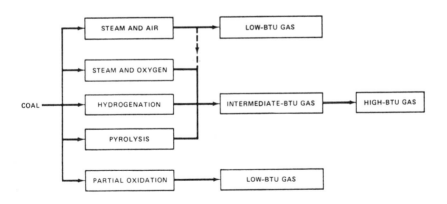

Source: U.S. DOE, *Energy Deskbook,* 1982.

removal of sulfur molecules, a major component of acid rain, and the primary environmental concern with coal combustion. The environment is not the only reason for removing sulfur in this process, as it "poisons" or causes rapid deterioration of the expensive catalysts used in turning the gasified coal into methanol. Its removal is an economic necessity for catalyst protection, but there also may be some cost offsets from marketing the sulfur.[40] Methanol's potential for removing this significant barrier to coal use has led the EPA to describe fuel methanol production as "the most promising use of coal."[41]

Currently, coal provides only 3% of the feedstock for methanol production. Only one plant in the U.S. produces methanol from coal.[42] Tennessee Eastman Co., located in Kingsport, Tennessee began producing acetic anhydride and other chemicals with methanol as an intermediate in 1984.[43] All plants have operated at greater than 90% capacity since the opening.[44] The facility produces 55 million gallons a year of methanol, all for chemical uses.

Just as methanol can provide a use for otherwise uneconomic supplies of remote natural gas, there are large amounts of low grade coal that are unmarketable because of their high-sulfur content or large amounts of chemically combined water. Certain types of lignite and peat may have up to 90% water and consequently little or no chance for use because of their low value and high transportation cost (they are still just as bulky). While this solid fuel may be too distant for electric power use by wire from coal steam generators, or too poor in value for use in boilers, often it is still suitable for gasification.

Economics of Current Methanol Production

Natural Gas

In order to meet current petrochemical and other specialized needs, it has been proved that gas-based methanol production facilities are economically viable. There are currently 16 such plants operating in the United States. However, if methanol is to become available in quantities large enough to supply a major part of the U.S. fuels markets, it is necessary to examine more closely the costs involved in the production and distribution of this commodity.

It takes about three years to construct a new methanol plant, which has a life of about fifteen years.[45] Costs involved are of two kinds: investment costs, which depend largely on location and plant size, and production costs which are linked mainly to the cost of the feedstock. Other costs include miscellaneous expenditures such as utility costs, insurance, general administration costs, selling and overhead. The return on investment is generally expected to be 20% before taxes.[46]

In September 1986 Jack Faucett Associates, an energy consulting group, prepared a report for the Environmental Protection Agency entitled, "Methanol Prices During Transition," which surveyed the costs for the current major methanol producers. The following discussion is drawn from the results of that report.

Capital costs for a methanol plant depend primarily on size and location. Plant size is related to efficiency, and usually indicated in tons/day. In the early 1970s 500 tons/day was considered large. However, today 1,500-2,000 tons/day plants are common. A plant of 227.5 million gallons/yr. (approximately equivalent to 1,500-2000 tons/day) was found to cost between $200-300 million depending on its location and the local support infrastructure.[47] The lower estimate would apply to centrally located plants in the U.S. or Canada while a similar plant in Alaska would involve higher costs. Less developed countries would require a higher ($300 million) investment because of a weaker infrastructure.[48]

Capital costs have been estimated to account for about 40% of the cost of gas-based methanol. Of this amount 40-60% is used in the preparation of the synthesis gas, the "reforming" step. Another 10-15% of the capital cost goes toward the distillation of the crude methanol.[49] Finally, the catalyst cost for natural gas based methanol averages about 1 cent a gallon of the product methanol.[50] A 275 million gallon/year capacity methanol plant consumes 78 MM scf/day of natural gas.[51] The cost of the feedstock is 5-15% of the delivered cost of methanol.

A modern methanol plant is a highly instrumented and automated facility and therefore labor costs are generally low. While labor costs would vary according to location, the capacity utilization should have little bearing on this cost. The Jack Faucett Associates report estimated the labor cost at 1.5 cents per gallon.

In addition to daily maintenance, independent contractors would be required to perform maintenance checks every 12 to 18 months, and in case of unscheduled shutdowns. This cost would increase with plant size and in developing countries where capital costs are higher. Faucett Associates calculated an average of 3 cents a gallon maintenance cost.

Coal

While the cost of the feedstock is significantly cheaper, methanol from coal is considerably more expensive than gas-based methanol because of very high capital costs. Capital costs can comprise up to 70-

75% of the cost of the product methanol, and are about 2 1/2 times those for a gas-based plant. While this obviously signifies a high cost for the product, it also means a large portion of the cost is not subject to inflation due to capital already invested. A coal to methanol plant in Southern Illinois, processing over 25,000 tons/day, involves a total of $838 million in capital costs. Of this sum about 83% of the cost is in cleaning the synthesis gas, removing impurities and sulfur. The remaining 17% is involved in the actual production for the fuel.[52] The high capital costs have led the Department of Energy to estimate the cost of plant construction in the lower 48 states of $2 billion. This produces product methanol at a cost of $1.25 to $1.30 a gallon, and a gasoline-equivalent energy content would cost about $2.17 a gallon.[53]

Transportation

Transportation of methanol is currently handled by special chemical tankers. These are generally much smaller (20,000 - 40,000 dead weight tons)(dwt), than crude oil tankers (up to over 300,000 dwt). Odfjell Westfol-Larsen, one of the two largest tanker operators in the world currently transports over 300 million gallons of methanol a year, with no problems. Tanks for methanol would be lined in either zinc-silicate or stainless steel, and fire detection and suppression systems would be necessary at an additional 1-3% of the cost of the production of the tanker.[54]

The potential for methanol transport is aptly summarized in the Faucett Associates Report:

> Most of the current liquid transport experience is with crude oil and oil products but methanol could easily become the second largest (by volume) liquid product transported and could surpass petroleum in some scenarios by early next century . . . . In the long run if methanol demand reaches levels of 20-30% US transportation demand shipping rates will be similar to those currently charged for large movements of oil in dedicated tankers. Such rates would include the added cost for special equipment required for methanol.

Back to the Future - Biomass

Methanol was originally made from biomass (wood) and many currently see its synthetic production as the logical and beneficial use for today's unused biomass. Biomass is defined as anything derived directly or indirectly from plant photosynthesis. Primary sources include forestry and wood-processing residues, crop waste, animal wastes and energy crops. Secondary sources, such as urban waste, are a potential source of feedstock as well.[55] Approximately 2 billion people worldwide rely on biomass as a main source of energy.[56]

Wood is the most widely used of biomass energy supplies. Over half the wood cut every year is burned to produce energy. According to United Nations' statistics the world's largest fuel wood producers in order are: India, Brazil, China, Indonesia, the United States and Nigeria. Currently about 2/3 of the wood used for energy in the U.S. goes into industrial, commercial and utility applications.[57]

The International Energy Agency has found the energy potential from forest and timber residues to be the highest in the U.S. with over 25% of the wood entering the timber industry available for conversion to energy. Commercial forests cover approximately 23% (2.1 million square kilometers) of the land area of the U.S.[58] This represents a potentially significant feedstock for methanol production. Another source is solid urban waste. The U.S. produces about 180 million metric tons of solid refuse per year. The energy in gas from this refuse is estimated to be about 2% of U.S. annual energy consumption. If converted to methanol it could supply about 8% of the fuel for our transportation needs.[59]

Similar to its production from coal or natural gas, methanol from biomass is produced by first gasifying the biomass and then manipulating the carbon:hydrogen ratio of the synthesis gas. The first modern biomass to methanol project was initiated at the Solar Energy Research Institute (SERI) in 1980, with the development of a one ton/day high pressure oxygen gasifier whose synthesis gas was then converted to methanol.[60] In Waltham, Massachusetts, Evergreen Energy Corporation formed a joint venture with Texaco adapting Texaco's high pressure coal gasification process with 55% efficiency, that accepts 3,500 tons/day of green wood chips to produce 330,000 gallons a day of methanol, an amount capable of providing up to 3% of New England's motor fuel when blended with gasoline. Total capital costs were estimated at $250 million, (1981 dollars), giving a product price of 80 cents a gallon.[61] In Richland, Washington, Batelle Pacific NW Laboratories conceptualized a $146 million plant that would process 1,800 metric tons per day (mtd) of wood to produce 900 mtd methanol (350,000 gallons/day) at 69 cents a gallon (1981 dollars).[62] The possibility of producing methanol from peat reserves was investigated in North Carolina where a 64 million gallons/year (175,000 gallons day) plant is being planned at a cost of $210 million (1981 dollars). The plant will use 2,000 mtd of peat to produce methanol at a cost of 75 cents a gallon.[63]

Many cities, such as Seattle, have conducted a detailed design study on the prospects of converting their municipal solid waste to methanol. Perhaps the most detailed examination of such a project was prepared for New York State's Energy Research and Development Authority. Because nearly all the fuels consumed in New York are imported from far away (U.S. Gulf and foreign sources) their expense is substantial and the state's vulnerability to external factors is high. In 1983 NYSERDA commissioned an investigation of producing alternative liquid fuels from New York State's indigenous resources. A detailed analysis of seven indigenous resources and eleven conceptual plants came to the following conclusions:

1. The production of methanol using small scale plants is feasible but not yet economically competitive at current market prices (60 cents a gallon)

2. In the long term methanol fuel produced from several indigenous resources (most likely forest biomass, peat, landfill gas and unconventional shale gas) has the potential to contribute up to 7,000 TPD (693 million gallons/yr) in liquid fuel supplies for New York State.[64]

The production of methanol from biomass has two major disadvantages which keep it from economic viability for the present. The first and biggest impediment to economic feasibility is the collection of the feedstock. Because biomass sources are dispersed over wide areas and are bulky to handle, their collection and transportation costs are high. This is especially true of waste. Furthermore, biomass generally has less energy per pound than other feedstocks, including lower ranked coals or lignite. These two characteristics generally make it more expensive for methanol production. The low energy content of the feedstock usually requires large production facilities to be economic, yet the dispersed nature of the feedstock, along with high collection and transportation costs, implies decentralization.

To combat these problems smaller, mobile production modules have been investigated by the farm equipment firm, International Harvester. To take advantage of smaller collection areas, IH has developed smaller, modular packaged plants (6 million gallon/yr, 50MTD) that are mass produced and delivered by truck or rail to operation sites. The mass production of identical units has been estimated to lower the capital costs enough to produce methanol for the same price as that manufactured at a plant fifteen times its size.[65] Furthermore, the plants could be moved as needed, closer to new sources of biomass feedstock, giving the plant a 20% higher load factor, and a return on capital investment that is three to four times that of a stationary plant.[66] Although this technology is not yet commercial, it indicates the potential of this source of energy supply and the serious consideration it is being given.

Seafuel[R] Synthesis Process: Methanol from the Oceans

There is a process that is uniquely designed to synthesize methanol making maximum use of renewable energy sources and without using petroleum, coal, naptha or natural gas. It is called the Seafuel[R] Synthesis Process or, SSP. The patented SSP process is shown in Figure 6-2.

The primary natural source for hydrogen is water and the SSP method derives hydrogen from seawater by the very well known methods of reverse osmosis and electrolysis, hence the name Seafuel[R]. The other major chemical component of the process is carbon dioxide (CO_2); and the sea also is one major source of CO_2, where it is dissolved in the water. The SSP process requires a low-cost supply of electricity and a similarly low-cost supply of CO_2. Several renewable sources of electric power are shown on the SSP diagram, along with various natural sources of carbon dioxide.

The core of the SSP process is the mixing of the elemental hydrogen and carbon dioxide, and reacting them in the presence of a catalyst to produce methanol and water. The methanol and water mixture then moves through a simple distillation process, which separates the two liquids. The effect, therefore, is that where there was once nothing but sunlight, wind and some machinery, there is now a storable, portable liquid fuel--methanol.

Note that except for the manufacture of some construction materials, the depicted process does not have to be dependent on fossil

Figure 6-2

SEAFUEL SYNTHESIS PROCESS
SSP

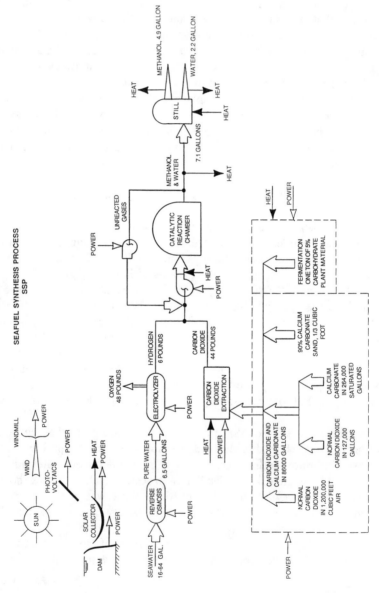

Source: Perry Energy Systems, Inc.

fuels (coal, natural gas, petroleum). There is no hydro-carbon feedstock needed for a synthesis gas phase of production. The electricity would, necessarily, be from the lowest-cost source available at the time and location of production; and several non-fossil energy sources are available, namely, geothermal, solar, wind, nuclear, primary batteries and, in the future, fusion. Thus, the choice of energy source is dependent on the availability at a given site and time.

The sources of hydrogen and carbon dioxide are also site and time dependent. For example, if fresh or pure water is available, reverse osmosis is not required. In addition, while carbon dioxide is universally available in air, the content of carbon dioxide in seawater is dependent on the water temperature and the availability of calcium carbonate sands. Sources such as the 5% carbohydrate plant material are wholly dependent on geography. The data depicted for carbon dioxide and calcium carbonate in seawater and calcium carbonate sands are specific to tropical islands. In cold northern waters, the availability of CO_2 through the calcium carbonate sands and calcium carbonate content of seawater is low; however, the normal carbon dioxide in cold seawater is 50% greater than that for tropical seawater.

A technically ideal site for initial SSP methanol production is the tropical island with its limited energy needs and the year round availability of plentiful solar and wind energy, seawater and calcium carbonate. Here the methanol can be produced and stored almost continuously and then either used indigenously or transported when and where needed. This site specific dependency is not unlike other existing methanol synthesis processes which make use of other waste or byproduct streams or that produce byproducts of value to other processes. It is likely that such decentralized facilities would be the first applications that are economic, with a specialized set of circumstances where the costs of fuel and other energy production costs are high.

The core of SSP, however, is not site specific. It only requires hydrogen, carbon dioxide, power and heat inputs to synthesize methanol; and, unlike other methanol synthesis processes, it operates at relatively low pressure and temperature. Any plant or process that produces any of the required SSP inputs as waste or excess outputs can be synergistically coupled to the core of SSP to produce methanol.

One example, the chlor-alkali industry, uses salt, water, heat and power inputs to produce caustic soda and chlorine plus an excess of hydrogen and heat. The excess hydrogen is approximately two pounds for each 150 pounds of caustic soda plus chlorine. In 1979 caustic soda and chlorine were, respectively, the seventh and eighth highest volume chemicals produced in the United States. Thus, the chlor-alkali industry may offer a large potential for synthesizing methanol at low cost with the SSP process. Petroleum refining, and ammonia and syngas production, also produce significant quantities of hydrogen and heat which, under special circumstances, may be better used to synthesize methanol.

Similarly, carbon dioxide is a component of waste and low value streams from many other processes: fuel oil, gas and coke combustion flue gases, 10-18%; lime and cement production gases, 10-40%; gases from the reproduction of starches, sugar and alcohol by fermentation, 99%. Excess carbon dioxide is also produced during the manufacture of ammonia and syngas from coal, and from petroleum and natural gas feedstocks. Methods for recovery of carbon dioxide from all of these processes are well known, and the recovery process also produces usable heat.

Ethanol (ethyl alcohol) is synthesized from petroleum and natural gas feedstocks. Fermentation of the carbohydrate content of plant materials produces the ethyl alcohol content of alcoholic beverages. Because ethanol can also be used as a fuel, agricultural interests propose and have constructed fuel ethanol facilities based on a fermentation of high carbohydrate content, grains such as corn, wheat, rye, milo, rice, etc. These grains have up to 75% carbohydrate content by weight. Such fermentation facilities present a unique opportunity for integrating SSP as follows:

o The fermentation process produces enough carbon dioxide to synthesize 7 gallons of methanol for each 10 gallons of ethanol.

o It also produces heat which can be used to reduce the electrical power needed to electrolyze hydrogen from water (See subsequent discussion).

o As a result, 50 pounds of 50% carbohydrate plant material through an integrated SSP-ethanol fermentation plant can produce a methanol-ethanol fuel mix with an equivalent energy content of 40 Kw-hours vs 25 Kw-hours if only ethanol was produced.

Another cost-saving approach to obtaining SSP feedstocks is by reacting carbon with the process' excess oxygen. Pure carbon is produced for a wide range of industrial purposes: activated for decolorizing and solvent recovery; carbon black for rubber and printing inks; carbon filaments and whiskers for composite materials, etc. When carbon is reacted with oxygen (oxidized or burned) the product is carbon dioxide. The SSP diagram shows that the electrolysis of water to produce hydrogen also produces about 10 pounds of oxygen per gallon of methanol. Two thirds of this oxygen can be reacted with 2.5 pounds of carbon per gallon of methanol to produce the carbon dioxide for the SSP process, plus usable heat equivalent to 10 Kwh per gallon of methanol.

Thus far, the potential applicability of the SSP process using naturally occurring resources, and its integration with man made hydrogen and carbon dioxide resources, have been discussed. This is not to imply that man made resources discard hydrogen and carbon dioxide. Most of the processes identified here recover these gases, either because of environmental concerns or because of their market value. What is implied, however, is that for new facilities and even for existing plants, it may be cost effective to integrate them with SSP and produce methanol from their waste and by-product streams.

The applicability of SSP in a particular circumstance is dependent on its cost effectiveness. For the previously discussed tropical island, cost effectiveness will be based on the cost of producing methanol in comparison to the cost of delivering an equivalent amount of energy to the island. For integration with other industrial processes, cost effectiveness will be defined by the following factors:

o The improvement of return on investment, i.e., can the return on an existing or new facility be improved by integration with SSP?

o The reduction of the cost of the inputs to SSP, i.e., can the SSP methanol synthesis cost be reduced by utilizing waste and by-products streams of other processes?

Both of these factors can be used to examine the potential cost effectiveness of SSP integration with other energy producing facilities.

The SSP diagram includes renewable energy sources. If these were replaced by purchased electrical power, the operating cost per gallon of methanol would be strongly dependent on the cost of the power (cents per Kwh). About 60% of the electrical power would be consumed by the electrolysis, approximately 30% for obtaining the carbon dioxide, and the remainder for the SSP core reaction processes. These, of course, are estimated relationships and, as previously discussed, they would be strongly site and time dependent.

Generally, hydroelectric plants provide the lowest cost electrical power. However, low-cost power is in high demand, and it will be difficult to find substantial amounts of surplus power from these sources. There are, however, plants that have been constructed in regions where for the next several years, there will be an excess of generating capacity to the region's needs. In some cases, it may be possible to identify such regions and put a deal together that would provide low-cost power for SSP that might not otherwise be marketable. This could improve the return on investment for the original plant. Furthermore, for those power plants needing additional peak power over and above their design capacity, the methanol could provide a version of electric storage, being used when needed to fuel a combustion turbine.

As previously indicated, approximately 60% of the energy for SSP is consumed in electrolyzing water to produce hydrogen. The energy excites the water's hydrogen and oxygen atoms which causes that bond between them to break, i.e., splits the water molecule into its component elements. In the technical world, energy comes in two forms, heat and the ability to do work. About 15% of the water-splitting energy can be supplied by heat, the remainder has to be provided by work. Work can be provided by electricity or radiation (nuclear or light).

The above discussion leads to an SSP integration that reduces the cost of the hydrogen input to SSP. If an SSP were integrated with a nuclear power plant, as shown in Figure 6-3, excess heat and radiation could be provided for the thermochemical reaction and the photolysis. Note that unused electric capacity of the nuclear plant could also provide the SSP electrical power. As previously discussed, the latter also improves the return on the plant investment. Further note that the nuclear plant element can be replaced by solar collectors. Last, but not least, if the nuclear plant were located near a source of low cost, pipeline carbon dioxide, which is a natural resource associated with oil wells, the cost of this SSP input could be reduced.

A fuel-fired power plant can be integrated to both improve the return on plant investment and to reduce the cost of SSP inputs. This is shown on Figure 6-4. In this, CO_2 is scrubbed from the plant flue gases, excess heat is used for thermochemical reactions and the SSP distillation process, and the unused electrical power capacity is used for electrolysis and the methanol synthesis.

Figure 6-3

Methanol Synthesis Integrated with Nuclear Power Plant

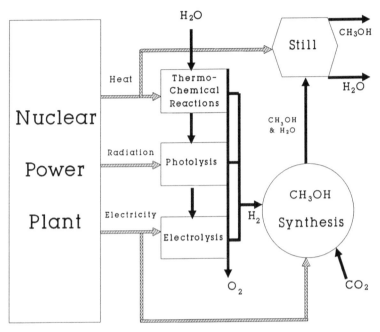

Figure 6-4

Methanol Synthesis Integrated with Fuel-Fired Power Plant

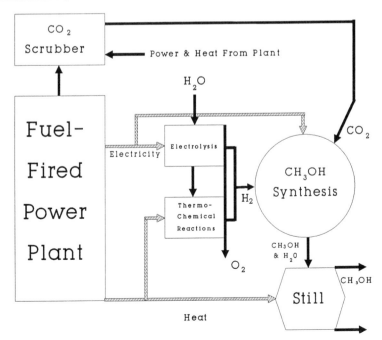

What has been presented is that SSP can derive methanol from renewable energies and natural, non-hydrocarbon sources (Figure 6-2), that SSP can be technically integrated with various processes and power plants (figures 6-3 and 6-4) and that, depending on site conditions, these SSP uses could be cost effective.

Hydrocarbon (coal, oil, natural gas) fuel sources will deplete in time. As they become scarcer, the cost to the general public will increase dramatically. Between now and the energy sources of the future, possibly fusion and hydrogen, methanol can be a bridge. An early start in constructing that bridge can help reduce the potential future trauma of hydrocarbon depletion.

Chapter Six: Summary

Methanol has been manufactured for over two hundred years. Originally produced by heating wood in the absence of air, methanol was first manufactured synthetically in 1923 from coal. Today natural gas is the feedstock for 90% of all methanol produced because of its higher hydrogen content and low feedstock costs. Ninety percent of all methanol produced today is used for chemical purposes. Of the small but growing amount used in the transportation sector, 95% is used for making MTBE, a fuel additive for unleaded gasoline that increases octane.

The chemical methanol market in the U.S. has undergone wild fluctuations of supply and demand in the past five years, including an oversupply in 1985-86 that sent prices plunging to 26 cents a gallon. While prices such as these would make methanol a very attractive motor fuel, it has had a negative impact on the industry, forcing shutdowns. Consequently, while the U.S. has sufficient feedstock for methanol production to fill chemical needs, production capability needs to be drastically increased before even a fraction of the transportation sector's demands could be met. Total 1986 methanol production amounted to a quantity that is equivalent to about 60,000 barrels of crude oil a day, only about 0.06 of the energy consumption of the transportation sector.

Methanol's production from a variety of sources holds significance for the future. Over 20 billion cubic feet of gas are flared daily worldwide, an amount equivalent in energy value to over three million barrels of oil a day. Furthermore, the U.S. has enough methanol in its coal reserves to substitute for over 300 years of transportation fuel demand at today's level of gasoline consumption. While natural gas is currently the only economically viable feedstock for commercial methanol fuel production, this will change as processes become more efficient and petroleum prices increase.

Methanol's flexible feedstock base can offer substantial benefits to U.S. energy security, since it can be manufactured from feedstocks as diverse as coal, biomass and water. The Seafuel^R process offers a virtually unlimited feedstock, the oceans, which provide both the hydrogen and the carbon dioxide necessary for methanol manufacture. Small scale demonstrations have been successful, and the process may currently be economic for use in remote (island) applications. The economics of large scale production of methanol using the Seafuel^R process will be dependent on available supplies of low-cost electricity.

Chapter Six: Vital Statistics

First synthetic methanol plant operated in Germany, 1923.

Current U.S. methanol consumption: 90% for production of chemicals.

U.S. methanol production in 1986: 1.2 billion gallons, about 0.5% of 1986 energy use in the transportation sector.

More than 100 new plants needed to supply 1 MMB/D of methanol for transportation. Plant cost estimated to be $200-300 million using natural gas feedstock.

Estimated 20 billion cubic feet of natural gas flared daily.

Methanol production and transport is much cheaper than LNG production and transport.

The U.S. has 20 times more energy stored in coal than the Middle East has in oil.

Commercial forests cover about 23% of U.S. land area, providing a significant feedstock for methanol production.

Seawater can provide all components of methanol: hydrogen; oxygen; carbon (from dissolved CO_2).

About 60% of the energy for seafuel production is consumed in electrolysis to produce hydrogen and oxygen.

Chapter Six: End Notes

[1]Marsden, S.S. Jr. *Annual Review of Energy,* 1983, p. 335.
[2]*Methanol Technology and Application in Motor Fuels,* ed. J.K. Paul, 1978. p. 18.
[3]*Ibid.*
[4]Jack Faucett Associates, "Methanol Prices During Transition" September 12, 1986, p. 3.
[5]Marsden, *op. cit.,* p. 337.
[6]Jack Faucett Associates, *op. cit.*
[7]Fossil fuels must be purged of all sulfur compounds to protect the synthesis catalyst.
[8]Long, F.W., *U.S. Petrochemicals,* 1972, p. 2
[9]Fuel grade methanol needs less distillation than chemical grade methanol.
[10]Long, *op. cit.*
[11]*Ibid.*
[12]Haggin, J. "World Methanol Situation Poses Challenge in Process Design", *Chemical and Engineering News,* July 16, 1984, p. 34.
[13]Jack Faucett Associates, *op. cit.*
[14]Information Resources, Inc., *op. cit.,* p. 163.
[15]*Ibid.*
[16]DOE/PE, *op. cit.,* p. 20
[17]Dixon, R.H., Product Manager, Gantrade Corporation "Methanol: A U.S. and Global Picture," Oxygenated Fuel Conference, Arlington, VA, November 18, 1987.
[18]Begley, Ronald, "Crude Rise Nudges Olefins Feeds: Petrochemicals '87," *Chemical Marketing Reporter,* April 6, 1987, p. 30.
[19]Agoos, Alice, with Kristine Portnoy, "Methanol Has a New Look; Tight Supply," *Chemical Week,* November 18, 1987, p. 44.
[20]Dixon, *op. cit.*
[21]Information Resources, Inc. *op. cit.,* p. 164.
[22]DOE/PE, *op. cit.,* p. 36.
[23]Agoos, *Chemical Week,* November 18, 1987, p. 48.
[24]Information Resources, Inc., *op. cit.,*
[25]"U.S. Fuel Alcohol Demand to Increase Thru 1980." *Alcohol Week,* November 9, 1987, p. 9.
[26]"Estimate Cut for U.S. Potential Gas Resources," *Oil and Gas Journal,* April 27, 1987, p. 105.

174 METHANOL: Bridge to a Renewable Energy Future

[27]O'Hare, T.E., R.S. Sapienza, D. Mahajan and G.T. Skaperdas, "Methanol for Transportation of Natural Gas Values," presented at the American Society of Mechanical Engineers Joint Methanol Conference, June 25-27, 1986, p. 1.

[28]Marsden, *op. cit.,* p. 338.

[29]Stinson, Steven C., "Methanol Primed for Future Energy Role," *Chemical and Engineering News,* April 2, 1979, p. 29.

[30]Roeder Bill, "Piping Methanol from Alaska," *Newsweek,* May 18, 1981, p. 37.

[31]Wentworth, Othmer, *Chemical Engineering Progress,* August 1982, p. 30.

[32]*Energy Security,* p. 116.

[33]Wentworth, Othmer, *Chemical Engineering Progress,* August 1982, p. 30.

[34]Marsden, *op. cit.,* p. 343.

[35]Jack Faucett Associates, *op. cit.*

[36]Berry, Bryan H. "Auto Energy: Meeting the Need for the Future," *Iron Age,* June 24, 1983, p. 43.

[37]Pope, Christopher, "Methanol Future Planned," *Renewable Energy News,* June 1984, p. 6.

[38]Reed and Lerner, *Science,* 1972, p. 1302.

[39]"The New Route from CO to methanol - via Water," *Chemical Week,* October 10, 1984, p. 50.

[40]*Alcohol Week,* February 13, 1984.

[41]Wilson, Dick, EPA Director of Office of Mobile Sources, testimony before Senate Commerce, Science, and Transportation Consumer Subcommittee, November 1987.

[42]*Alcohol Week,* "EPA Says Fuel Methanol Use Is Only Practical Way To Reduce Ozone Pollution," November 16, 1987, p. 3.

[43]An intermediate is a compound formed between the initial and final stages of a process or reaction.

[44]Statement of Robert E. Long, Director of Strategic Planning, Eastman Chemical Division, Eastman Kodak Co. before the Subcommittee on Fossil and Synthetic Fuels and Subcommittee on Energy Conservation and Power, April 25, 1984.

[45]Jack Faucett Associates, *op. cit.,* p. 51.

[46]*Ibid.,* p. 54.

[47]One metric ton is 2240 lbs. One gallon of methanol is 5.8 lbs. Therefore, there are 2240/5.8, or approximately 390 gallons of methanol in one metric ton.

[48]Jack Faucett Associates, p. 51. Countries mentioned were Argentina, Brazil, Chile, Trinidad, Burma, China, India, and Malaysia.

[49]Haggin, J., July 16, 1984. p. 32.

[50]Jack Faucett Associates, p. 35.

[51]Cohen, Lawrence H. and Herman L. Muller, "Methanol Cannot Economically Dislodge Gasoline," *Oil and Gas Journal,* January 28, 1985, p. 122.

[52]*Ibid.,* p. 123.

[53]March 1988 draft of Information Resources, Inc. report. Not in September 1988 version.

[54]Jack Faucett Associates, *op. cit.,* p. 46.

[55]Shea, Cynthia Pollock, "Renewable Energy: Today's Contribution, Tomorrow's Promise," Worldwatch Paper no. 81, January 1988. p. 18.

[56]Deudney, Flavin, *Renewable Energy, The Power to Choose,* p. 108.

[57]Shea, *op. cit.,* p. 19.

[58]*Ibid.*

[59]Reed & Lerner, p. 1303.

[60]Department of Energy Activities on Methanol Report, prepared for the House Subcommittee on Fossil and Synthetic Fuels, April 2, 1984.

[61]Haggin, Joseph, "Methanol From Biomass Draws Closer to Market," *Chemical & Engineering News,* July 12, 1982, p. 24.

[62]*Ibid.*

[63]*Ibid.*

[64]Wan, Edward I., Joseph D. Price, John A. Simmons, "Methanol Production from Indigenous Resources in New York State: Vol. I Executive Summary," p. 3-1.

[65]Marsden, S.S., Jr., 1983, p. 344.

[66]Haggin, J., July 12, 1982, p. 24.

CHAPTER SEVEN

A BRIDGE TO THE FUTURE

The fuel cell is the first truly new generating technology to come along since nuclear.

-Arnold Fickett

The focus of this book so far has been to: examine the strengths and weaknesses of the current fossil fuel-based energy economy in the U.S.; evaluate the near-term alternatives to fossil fuels; and take a careful look at methanol as a fuel whose time has come, primarily in the transportation sector. The key options to petroleum that have been examined for U.S. transportation use include:

o A much costlier, less powerful electric vehicle fleet, focused initially on "citicars" due to range limitations and problems with battery technology and recharging times;

o A compressed natural gas (CNG) fleet that requires substantial capital investment in both vehicles and a complex refueling infrastructure, and is likely to focus on the niche of larger fleet vehicle applications because of weight trade-offs;

o Ethanol as a neat alcohol fuel, which fits best with the current vehicle design and fuel distribution system but is the most expensive of current liquid fuel alternatives;

o Methanol as an automobile and heavy vehicle fuel, which is cost-competitive and fits well into the existing gasoline-powered vehicle infrastructure but requires some capital investment for both vehicle manufacture and refueling.

Chapters 4 through 6 have examined in detail the applicability of methanol to U.S. transportation markets, the problems that are faced in penetrating those markets, and the options for methanol production at levels to meet transportation needs in the U.S. In this chapter we will examine the more distant future for energy markets in the U.S. and how methanol may play a key role as the bridge to a quite different renewable energy future.

There is growing agreement among futurists that the ultimate renewable energy economy would be fueled by hydrogen. The simplest and most abundant element in the universe, hydrogen accounts for 70 to 80% of the visible universe. It is the lightest of the elements with an atomic weight of 1. Hydrogen accounts for 11.9% of water and is an essential constituent of all acids, hydrocarbons and vegetable and animal matter. The sun, which accounts directly and indirectly for most of the energy consumed on our planet, consists of the continual fusion of hydrogen atoms to produce helium.

Hydrogen's physical and chemical properties have long made it every energy thinker's dream fuel. Its thermal efficiency is very high; and on the basis of weight, it gives about three times as much heat per pound as gasoline. Hydrogen has been described as "the most efficient known method of storing energy."[1] Because pure hydrogen contains none of the carbon found in fossil fuels, it is the environmentalists' dream as well. In combustion, its only by-product is steam (and significant amounts of heat-induced nitrogen oxides).

Although hydrogen is our most abundant element, it is highly reactive and does not exist freely in nature like coal or petroleum; and it cannot be drilled for or mined from the earth. Its abundance is due to its presence everywhere, but it must be manufactured from other compounds.

There are two main sources of hydrogen: water and hydrocarbons. Although almost a dozen processes exist for the commercial production of hydrogen, most rely on the extraction of hydrogen from hydrocarbons. Similar to the process used to obtain synthesis gas from coal or natural gas, "steam reforming" involves the reaction of hydrocarbons (usually natural gas) with steam at high temperatures ($850^{\circ}C$).[2] Electrolysis, where water molecules are broken into elemental hydrogen and oxygen, produces commercial volumes of hydrogen by running an electric current through water. However, this process requires large amounts of electricity (94 kwh per Mcf of hydrogen); and it is used mainly where electricity is cheap and a high purity of hydrogen is required.[3]

The necessity for low-cost electricity is a limiting factor in the production of hydrogen in general, where the cost of energy represents 70-80% of the production cost.[4] The high temperatures required for steam reforming, low overall yields, and expensive process materials make even this approach to hydrogen production uneconomic in today's fuel markets. Currently hydrogen is sold at several times the price of petroleum. A hydrogen authority at Texas A&M cites projected costs for hydrogen in the range of $45/million BTUs in the mid-1990s, equivalent to gasoline at about $6/gallon; but new technologies for producing hydrogen from coal could produce hydrogen for about $9/million BTUs, equivalent to gasoline at $1.20/gallon. Hydrogen production via electrolysis is even more expensive than steam reforming because the former depends almost solely on the cost of electricity, which can run more than four times the cost of natural gas.[5] For this reason hydrogen supporters have focused on the means of reducing the high cost of electricity, including use of "cheap" nuclear power or some less expensive form of solar energy.

Peter Hoffmann, a hydrogen enthusiast and author of The Forever Fuel; The Story of Hydrogen, addresses the broader costs of hydrogen production and use: "There are many technical problems in the storage, handling, and safe use of hydrogen, but almost everybody agrees that once hydrogen can be produced at prices competitive with fossil fuel, half the battle would be won."[6] In this statement lies an essential barrier to hydrogen's adoption as a fuel today. Cost of production is only half the battle, as storage, handling, and safety still need to be addressed. The problems with hydrogen in existing automobiles were addressed in some depth in Hoffmann's account:

With hydrogen, on board fuel storage represents one of the key obstacles to the automotive "clean machine". The solution most people think of--storing the gas in steel-walled pressure tanks similar to those used in welding--is out because of excessive weight, although it poses the least design problems. One researcher . . . calculated once that a hydrogen pressure tank capable of holding roughly the same amount of energy as an 80 liter (17.6 gallon) fuel tank for a standard sized car would have to weigh about 1.5 metric tons (1.7 short tons) and would require a pressure of 800 atmospheres as well as steel wall thickness of 7 cm. (almost 3 inches). Exotic materials such as titanium, carbon, boron, glass composites or cryogenically formed steels could cut the weight by a maximum factor of about five to maybe 300 kilograms (660 lbs)-- still far too heavy. Even if a pressurized-gas container of acceptable weight, comparable to the 50 kilograms of a conventional gas tank could be developed, pressure containers represent a serious hazard in case of a collision. "These tanks

*would explode as if they contained high explosives, causing great
blast damage and greatly increasing the danger of fire."*[7]

As a consequence, there are two main methods currently being
considered for the storage of hydrogen. The first of these is via
cryogenics, or the storage of hydrogen at very low temperatures (-252° C
or below). While this method has certain logistical advantages, (refueling
time, distribution and relatively fewer safety problems), the economics are
unattractive. Hydrogen liquefaction is very expensive, and a fuel tank on
board a conventional automobile would cost about $2,000 for storing the
same energy content as in a conventional car. Finally, fueling systems
would have to be retrofitted entirely to meet the needs of hydrogen
supply.

The second means of storing hydrogen is via a hydride, a
compound of hydrogen and another element, that looks and feels like a
metal but absorbs hydrogen like a sponge by bonding the hydrogen atom
atomically. Heat is added to the hydride to release the hydrogen for
combustion. While this method would take up no additional volume on
board (of course cars would have to be designed for its use), the storage
equivalent to a 100 liter (26.4 gallon) "tank" of liquid hydrogen would
weigh 25 times as much as a 100 liter metal tank.[8]

If hydrogen were cheaper than fossil fuels and its use could be
implemented within the existing infrastructure, there is still no guarantee
that it would be implemented in the transportation sector. The
limitations of methanol's entry into the transportation sector bear
testament to this. But because hydrogen would necessitate such dramatic
changes in fuel transportation, storage and use technologies, and because
it is more expensive and more difficult to handle, its adoption based
strictly on environmental appeal is difficult to justify. In 1974 at The
Hydrogen Economy Miami Energy (THEME) conference, a group of
researchers from Stanford presented a paper on the relative merits and
inconveniences of hydrogen as an automotive fuel, concluding:

*Compared to several alternatives, especially the use of methanol in
vehicles, a transition to hydrogen would appear to be needlessly
disruptive . . . it is our present conclusion that it is very unlikely that
the transportation system will evolve of its own accord in the
direction using hydrogen as a fuel for private vehicles; Moreover,
government intervention to alter this state of affairs in favor of a
hydrogen system is unlikely to be warranted.*[9]

The numerous limitations on hydrogen's use as a transportation fuel have led many to the conclusion that in the medium term, some fuel other than hydrogen will be needed to bridge the gap to renewables. Dr. T. Nejat Veziroglu, President of the International Association for Hydrogen Energy and Director of the Clean Energy Research Institute of the University of Miami, enumerated seven preconditions for an intermediate energy carrier in his support for bills before the U.S. Congress favoring hydrogen research. He noted: "The shortcoming of the renewable energy resources point out that the need for an intermediary energy system to form the link between the non-fossil energy resources and the user." The criteria identified for such a system are:

1. It must be storable
2. It must be transportable
3. It must be a fuel for transportation
4. It must be efficient
5. It must be environmentally compatible
6. It must be economical
7. It must be recyclable, if possible

While Dr. Veziroglu argues that hydrogen satisfies all of these criteria, methanol also fits the description, at a much reduced cost and far less inconvenience to the existing fuel system.

Storing Hydrogen in Methanol

Methanol's easier handling and lower production costs have already led even hydrogen advocates to suggest conversion to methanol as the ideal means of storing hydrogen. One study found that when natural gas is sold at the wellhead for between 60 and 80 cents/MMBTU, the lower investment and reduced energy consumption of methanol production makes hydrogen manufacture, via conversion to methanol, 20% cheaper than LNG, when the distance between producer and consumer is 6,500 miles or greater.[10]

The steam reforming of methanol to produce hydrogen is claimed to be competitive with electrolytic hydrogen (hydrogen produced via electrolysis). A 250 m^3/h pilot plant has been developed by a French and Belgian company who believes that this process will be competitive with electrolytic hydrogen because of lower temperatures required for reforming and no process heat required after start-up (since this heat can be generated by burning methanol).[11] While steam reforming may offer another alternative for the production of hydrogen, it does not offer a solution of the problems associated with its use in existing and near future automobiles. But in its similarity to hydrogen (and methanol is the fuel most similar to hydrogen in its environmental characteristics), liquid methanol offers to bridge the gap to an energy future fueled by hydrogen. Hoffmann observes:

> *Methanol decomposed into hydrogen and carbon monoxide with an assist from engine heat gives most of the advantages of a pure hydrogen engine and few of the drawbacks such as heavy hydride tanks. Tests, especially those by Nissan, showed a 20 to 40% increase in fuel economy and no need for catalytic converters: "Such a system may well be better than a straight hydrogen engine, running on pure hydrogen . . . a lot depends on what kind of hydrogen storage device will be developed . . . it appears that it will be difficult to find a better storage device for hydrogen than liquid methanol for automotive applications."[12]*

The Fuel Cell

Imagine an electric car whose battery needs no recharging; running on the components of water, hydrogen and oxygen, it needs only to be refilled like contemporary automobiles. Unlike contemporary automobiles, however, it emits nothing but steam from the tail pipe. Furthermore, when the power is not needed, it can be switched off with no loss of power. This is essentially how a fuel cell powered car would operate and the obvious reason for its popularity among automotive visionaries.

In 1842 a scientist named William Grove wrote to Michael Faraday at the British Royal Institution describing a "gas battery" he had produced by reacting hydrogen and oxygen on platinum electrodes dipped into dilute sulfuric acid.[13] While the applications have increased dramatically, the principle behind the fuel cell remains the same. Essentially the reverse of electrolysis, a fuel cell involves the combination of hydrogen and oxygen to form electricity and water. Hydrogen and oxygen are fed into opposite sides of a cell and separated by an electroplated barrier that contains a substance (referred to as an electrolytic substance) that conducts current. An electrode strips the hydrogen of its electrons, and the resultant hydrogen nuclei pass through the conducting acid to combine with the oxygen. The products of this reaction are electricity, water and heat.[14] The fuel cell operates much in the same way as storage batteries in flashlights, transforming chemical energy into electricity (see Figure 7-1). The major difference is that the chemicals are used outside the cell. Consequently, the power supply can be in continuous operation-- as long as the fuel and oxidant are fed to it, the fuel cell will generate electricity--while a battery ceases to operate once its stored chemical energy is depleted.

Fuel cells have been used to power manned space flights since the 1960s, but their use on land has until recently been limited by the high cost of production and their high weight to power ratio. Early demonstration fuel cells were constructed during the push for alternate fuels but never commercialized. Improvements in fuel cell technology, however, have spurred a growth in the use of fuel cell-powered utilities; and as of January 1986, fuel cell technology was already handling the power needs for several types of businesses, such as a laundry, a telephone switching station, a Howard Johnson's restaurant in the Baltimore area, a hilltop Sheraton in an L.A. suburb, and a swimming pool in Japan.[15]

Figure 7-1

Simple Fuel Cell

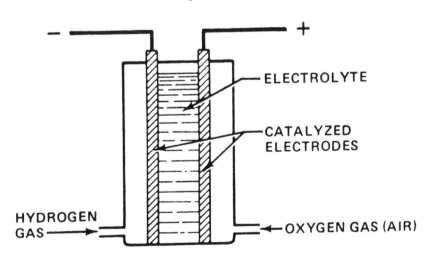

Fossil Fuel Cells

Original fuel cell technology used hydrogen gas and either oxygen gas or air. Fossil fuel cells use either a gaseous or liquid hydrocarbon as the source of hydrogen, and eventually even coal may be used as the hydrogen source. For oxygen, fossil fuel cells generally draw the amount needed from the air. There are three main components to cells that are designed to use fossil fuels (Figure 7-2). These are:

o The fuel processor, which converts the fossil fuel into a hydrogen-rich gas;

o The power section, which contains the cell or cells;

o The inverter which changes the direct current (DC) that the cell produces into alternating current (AC) to feed to the power grid.

Phosphoric acid cells are the most common, and that technology is near commercialization. Cell names (e.g., phosphoric acid, molten carbonate, etc.) are taken from the electrolyte that is used in the cell. The primary fuel used in most demonstrations has been natural gas or naptha, which is subjected to high-temperature steam reforming to get a catalytic reaction. The result is a gas mixture of hydrogen and carbon dioxide, which is supplied to the negative electrode. Air provides the oxygen for the positive side, and the cell's operating temperature is about 200° C.

Figure 7-2
Main Components of Fuel Cell Systems

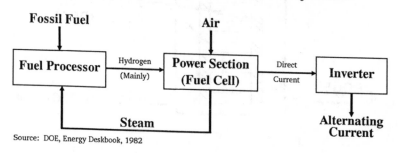

Source: DOE, Energy Deskbook, 1982

Today there are several dozen on-site 40 kilowatt fuel cells operating in the U.S., primarily phosphoric acid designs that are the result of a major demonstration program that began in 1978. The next phase includes plans to scale these models up to 200 kw capacity for use in apartment buildings, stores, light industry and similar sites.[16] In Japan, the leader in fuel cell technology, several utility-size fuel cells have been tested; and an 11 MW plant is scheduled to come on-line very soon.[17]

The idea of a fuel cell power plant is attractive for several reasons. Their most obvious advantage over oil, gas or coal-burning utilities is environmental. Direct chemical conversion eliminates the burning of fossil fuels, and along with it all traces of carbon or sulfur. But the advantage of direct chemical conversion goes beyond just the environmental implications. By avoiding the combustion of fuel, which requires energy input (the heat required to burn the oil, gas or coal), and with no need for mechanical energy (the electricity required to operate the boilers and turbines), efficiency is greatly improved. The U.S. average for electrical power generating efficiency is currently about 33%, which means that when electricity is generated, about 2/3 of the energy put into the generating process is given off as waste heat.[18] Since a fuel cell requires no energy input, the actual achieved efficiencies in electricity production exceed those of conventional power generation. A fuel cell using natural gas will produce electricity at a 40% efficiency of conversion, and it is estimated that efficiencies will approach 65% in the future. Coal-fueled fuel cell systems with comparable fuel cell technology are expected to reach efficiencies in the range of 35 to 60%.[19]

Further advantages of a fuel cell utility are its flexibility in construction, location, size and operation. The environmental and space requirements of conventional utilities place limits on their siting, usually requiring that they be located far from the centers for which most of their electricity is destined. High capital costs for new construction and expected economies of scale led historically to the construction of fewer and larger plants, rather than smaller increments of new generation. Over the past 10 to 15 years, this situation has changed. As was described in Chapter 2, sharp reductions in demand growth for electricity in the 1970s caused severe economic problems for electric utilities. Construction times were stretched-out several years, costs grew by as much as a factor of 10, and numerous plants were cancelled. The result has been an uneven distribution of generation capacity across the U.S., with several utilities that have found themselves holding large amounts of excess capacity while others are capacity-short. The current move is away from large capacity additions, a situation that is quite favorable for use of fuel cells.

Contrary to conventional views of power plant efficiency, fuel cell operating efficiency is relatively independent of power plant size. A 500 KWe power plant may have the same efficiency as a 500 MWe power plant, which means that smaller units of standardized modular construction can be added to a utility grid in a short period of time without sacrificing efficiency.[20] Shortened construction times reduce the problems of financing that accompany conventional utility construction, and thus directly benefit the rate payer who will only pay for the amount of electricity needed. Projections of eventual electricity costs using the less-expensive molten carbonate technology suggests that these natural gas-fueled systems could produce electricity for about 6 cents per kwh, which is quite competitive for new generation in today's market.[21]

Flexibility in the size of FC power plants, along with their environmental benefits, will have a significant impact on their location. Smaller increments of fuel cell generation capacity can be placed at different locations around the utilities' power grid, allowing the utilities to reduce the transmission costs as the plants are nearer to the electricity's end users. In congested urban centers, this flexibility is particularly useful, because transmission and distribution facilities are usually expensive to install and very difficult to get permitted. Furthermore, by locating fuel cells near points of utilization, the waste heat produced by the cell (the by-products of the cell are electricity, water and heat) can be captured and used for heating (space or water) or cooling purposes. Useful recovery of waste heat could result in fuel use systems with a combined energy efficiency of as much as 90%.[22]

Current impediments to increased use of hydrogen fuel cells are concerns with cost and longevity. Presently units are three to four times the price at which they would be competitive. Costs of component catalyst stacks (usually made of platinum) are high, as are the labor costs to produce them. Furthermore, the hydrogen and oxygen which fuel them must be manufactured from somewhere; and although feedstock costs are low (both can be produced from water), the cost of extraction is expensive. The overall costs will be reduced if the cells can be made to operate over long periods of time, such as five years, or 40,000 hours; but as of recently the longest running system had just passed 3,000 hours.[23]

It is easy to see why fuel cells are being seriously considered as a major element of our energy future. However, the problems of hydrogen storage and transportation would remain. One suggested solution has been the use of methanol as a storage medium for the hydrogen. A methanol-powered fuel cell works much the same way as the fuel cells described above. The only difference is that a reformer is added to the cell. Methanol is fed into the side designated for hydrogen, where it is reformed to produce hydrogen and carbon dioxide gases. The carbon dioxide is expelled and the hydrogen is used for the fuel cell. Reed and Lerner wrote in <u>Science</u>, "Methanol is one of few known fuels suited to power generation by fuel cells . . . Although methanol is not as simple to use in a fuel cell as hydrogen, it can be stored and shipped more easily."[24]

Today there are several kinds of fuel cells being produced, differing mainly in their chemical catalyst (and therefore the operating temperature), and power output, and their differences shape their respective applications. In the 1986 issue of Energy, the International Journal, an assessment of research needs for the advancement of fuel cells lists five primary types of fuel cells, whose relative merits and limitations are outlined below:

1. <u>Phosphoric Acid Fuel Cells</u> (PAFCs) - These fuel cells are the most mature, with twenty years development and $400 - $500 million dollars worth of research in them. There is a widespread belief in the U.S. that only PAFCs tolerate hydrocarbon fuels at lower temperatures, and so there has been greater support for these. As a result these cells are projected to establish a significant niche in electric and gas utility markets.

2. <u>Alkaline Fuel Cells</u> (AFCs) - These cells have powered all life support systems in Apollo spacecraft since the 1960s, but a lack of research efforts toward their terrestrial use leaves them more than ten years behind PAFCs. These cells offer better energy efficiency, higher power capability, lower operating temperature, better materials tolerance and overall better performance than PAFCs, and could possibly be used in electric vehicles someday; however, they require pure hydrogen or the cost effective removal of carbon dioxide from hydrocarbon fuels, which limits their feedstock.

3. <u>Solid Polymer Electrolyte Fuel Cells (SPEFCs)</u> - This was the first fuel cell used as a non-propulsive power plant for

manned Gemini Earth orbit missions (1963-65). Since then its use for terrestrial applications has been refined by General Electric. Disadvantages of this type fuel cell have included high electrolyte ($200/kw) and catalyst ($450/kw) costs, low tolerance to CO (<1ppm) and low operating temperatures (<100°C), which restricts the use of cell heat for fuel processing. Since 1984, Ballard Technology Corp., a Canadian battery manufacturer and fuel cell research group, has made significant technical advances in SPEFC design. These include lower catalyst loading, higher current densities, and alternate solid polymer membranes, all of which contribute to lower costs and improved performance. These advances enhance the potential that SPEFCs have for applications in the transportation sector.

4. <u>Molten Carbonate Fuel Cells</u> (MCFCs) - A relatively new concept, the MCFC is still about 7 to 9 years away from commercialization. Its advantages are its simple plant design, lower plant cost (relative to the PAFC), ability to accept CO, CO_2, and H_2 with high efficiency, and high quality of heat (550°C and higher) relative to total energy. Some see in MCFCs the potential for internal reforming of natural gas, but problems of endurance and performance must be solved before successful commercialization can occur.

5. <u>Solid Oxide Fuel Cells</u> - (SOFCs) - The development of SOFCs dates back to the mid 1950s but hasn't been intense until recently. In the past five years, Westinghouse has shown considerable interest in this type of cell, which is specific for technically sophisticated customers. The SOFC is distinct from other fuel cells in its high operating temperature (1000°C) which allows electrochemical oxidation of the hydrogen without an added catalyst, its fuel versatility and the constituency of its primary components-- ceramics. While the materials for this type of cell are cheap, their processing is very difficult; and its single largest need is in materials research.

Thus far, the fuel cells that are capable of running on methanol directly are too heavy for vehicular applications; and those which could be used must take pure hydrogen, for which problems of handling and storage remain. The cost is still prohibitive for commercial markets, but advances in fuel cell technology suggest that the day when automobiles could run on fuel cells may not be far away. Researchers at Argonne National Laboratory have made significant strides on the weight to power ratio by using advanced ceramics to construct lightweight, efficient fuel cells. Their design could be used to build a 100 kw cell, about 11 cubic feet, weighing approximately 300 lbs including cooling components (a typical internal combustion engine of comparable power, or approximately 132 horsepower, weighs 600 lbs).[25] Improvements like these will continue to bring us significantly closer to the use of fuel cells in automobiles, and a simpler system for converting methanol to hydrogen would be a technological breakthrough in the application of fuel cells for transportation.

The Stirling Engine

Another energy technology looming on the horizon which is capable of using hydrogen or methanol as a fuel source, is the Stirling engine. Capable of burning any fuel, highly efficient, and clean burning, it could make a significant contribution to the automotive industry with its high fuel economy. Stirling technology has had commercialization problems similar to those of fuel cells in that until recently, major drawbacks have been its weight, cost and unreliability. While these have previously prevented its serious consideration as an energy source for automotive applications, recent research efforts indicate its availability in that area may not be far away.

The concept of the Stirling engine is over 172 years old. It was first invented in 1816 by a British clergyman, and the key to Stirling technology is its external combustion.[26] Stirling technology is much different from conventional, open cycle, internal combustion engines, where vaporized fuel is injected into the cylinder, compressed, burned, and then combustion products ejected as exhaust. Stirling engines perform the power generation process in a closed cycle, where compression and expansion actions are achieved by alternatively heating and cooling the same working gas.

The three main characteristics that distinguish a Stirling engine from a spark-ignited open cycle engine are:

- the gas is contained in a continuous closed volume
- the volume is divided into hot/cold regions
- compression/expansion are accomplished by periodically varying the size of the volume[27]

A simple schematic for a double acting system is shown at Figure 7-3. This design, known as the Rinia Arrangement, contains two volumes of working gas. One flows from the top (expansion space) of cylinder A to the bottom (compression space) of cylinder B. The other flows in the reverse arrangement. The major components affecting each gas volume are the heater (H), the regenerator (R), and the cooler (C). An external heating source converts the energy in the fuel into a heat flux, heating the working gas (hydrogen) in the closed system. Two reciprocating pistons

Figure 7-3

Cylinder Pair in Rinia Arrangement of Stirling Engine

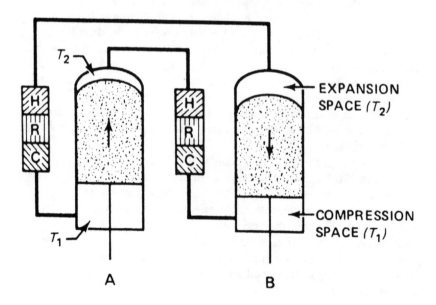

Source: DOE *Energy Deskbook*

remove and store the thermal energy from the hydrogen as it passes from the hot to cold region. The proper phasing of these processes transfer most of the gas in the cold region while compressing, and in the hot region while expanding, resulting in power generation. Finally a cold engine drive system transfers the motion of the piston to connecting rods, and a crankshaft converts the rotary motion into work.[28]

Since the combustion is external and only for the purpose of providing heat, the exhaust gases of the Stirling engine are recirculated, rather than expelled; and this both increases fuel economy and reduces harmful exhaust emissions. By recirculating exhaust gases through the combustor, the flame temperature is reduced, which in turn reduces the production of nitrogen oxide emissions. Consequently no catalytic converter is needed. Some modifications in automobile design are needed to accommodate a Stirling engine, including the relocation of the front cross member, a larger radiator and a second fan in front of the radiator.[29]

Until recently, the heavy weight of the engine (14 lb/hp), the high cost of manufacturing ($5,000), and the slow acceleration (0 to 60) in 36 seconds) were the principal deterrents for the application of the Stirling engine to automobiles; but recent successful research efforts have made the Stirling much more attractive.[30] The Automotive Stirling Engine Development Program, funded by the Department of Energy and administered by NASA at Mechanical Technology Incorporated, set out in 1978 to integrate the Stirling engine into an automobile within acceptable parameters of driveability. In a report published by NASA in 1986, the experiment was declared a success. Test results produced an 83.5 hp engine inside a 1985 Chevrolet Celebrity with the following results: after over 15,000 test hours, the Stirling engine in a Chevrolet Celebrity had achieved a combined fuel economy of 41 miles/gallon of gas, 50% better than the nation's fleet average. Acceleration time from 0-60 mph was 12.4 seconds while the conventional, spark ignition Chevrolet was 13.0 secs. With a weight of 5.5 lb/hp and a cost of $20/kw (or a manufacturing cost of $1,200), the Stirling engine is significantly closer to its automotive niche than it was ten years ago.

The Stirling engine still has to meet certain requirements before it will be accepted in the passenger car market. But its advantages have found potential applications in other markets, such as in submarine vessels, where the recycled exhaust would allow virtually undetectable movement (tracking systems are heat sensitive) and could enable a vessel to stay submerged for up to 14 days.[31] Furthermore, in a space station application a Stirling engine could draw upon solar heat, using almost no fuel.[32]

The concepts of the fuel cell and the Stirling engine are just two of many developing on the horizon of a future fueled by renewables, and each of them holds promise for an environmentally superior energy system unconstrained by limits of energy supply. Currently, however, many are limited by other factors, such as the relatively lower cost of alternatives (i.e. the existing system), a lack of research funding and continuity, and a relative resistance to change. Their entry into our energy system will be dictated by the urgency of our energy needs; and while the timing of this occasion is unpredictable, its inevitability is certain. The introduction of methanol into our transportation sector today can provide an easy transition to the unforeseeable but certain changes of tomorrow. As an environmentally benign energy source, one that can be used easily in today's transportation system while offering potential use in the systems of tomorrow, methanol offers a bridge over the gap between our current dependence on fossil fuels and the energy future beyond.

Chapter Seven: Summary

The notion of producing fuel for the nation's automobiles from seawater sounds as if were taken from a science fiction fantasy, but its realization only dramatizes methanol's potential to link a finite supply of fuel for today's automobiles with a tomorrow fueled by a renewable automotive fuel.

Hydrogen is commonly acknowledged as the optimal renewable energy source. The simplest and most abundant element in the universe, it is believed to be the original gas from which all life evolved. It is extremely efficient as a storage medium and very clean burning. But, as an energy source, it faces several problems of implementation, such as safety and handling; and its logistics of transportation and storage make a hydrogen fueled transport system economically impossible for the present.

Because of its high hydrogen content, methanol enjoys most of the benefits of hydrogen while its liquid form avoids most of the major problems associated with the storage and handling of gaseous hydrogen. The similarities in energy and environmental efficiency of methanol and hydrogen, combined with the crucial differences in form, have resulted in research and development on several energy technologies, such as the fuel cell and the Stirling engine, which could use methanol initially and evolve into hydrogen as its form is made more directly useful and economic in the future. In this way, methanol can bridge the gap between the fossil fuels of today and an environmentally clean, renewable energy future of tomorrow.

Chapter Seven: Vital Statistics

Cost of on board hydrogen storage for equivalent energy content of gasoline powered vehicles: $2,000

Factor by which weight of fuel storage is increased when using metal hydrides: 25

Energy efficiency of fuel conversion in electricity generation by conventional generators: about 1/3

Energy efficiency of fuel cell conversion to power:

>   Currently: 50%
>   In future: 65-90%

Fuel economy increase with Stirling engine: 50%

Cost of manufacture, assuming production run: $1,200

Chapter Seven: End Notes

[1]Hoffmann, Peter, "Fueling the Future With Hydrogen, *The Washington Post,* November 6, 1987, p. B-3.
[2]Jonchere, R., "Methanol Seen as Hydrogen Source," *Oil and Gas Journal,* June 14, 1976, p. 70.
[3]Hoffmann, Peter, *The Forever Fuel: The Story of Hydrogen,* 1981, p. 3.
[4]Jonchere, June 14, 1976, p. 72.
[5]Bern, Reginald I., "Hydrogen Routes' Future Is Keyed to Economics," *Chemical Engineer,* July 14, 1980, p. 80. (However, the Electric Power Research Institute in Palo Alto, California found that when hydrogen demand is under 100 million std ft^3/yr, capital costs are lower for electrolyzers in comparison with steam reformers, and for a 33 million std ft^3/yr capacity costs were \$5.51/1000 std ft^3 vs. \$8.54/100 std ft^3 for electrolysis and steam reforming respectively.
[6]Hoffmann, *op. cit.,* p. 52.
[7]*Ibid.,* pp. 117-118.
[8]*Ibid.,* p. 119.
[9]*Ibid.,* p. 131.
[10]Jonchere, *op. cit.,* p. 72.
[11]Bern, *Chemical Engineer,* July 14, 1986, p. 84.
[12]*Ibid.,* p. 135.
[13]Johnson Matthey, *Newsline,* January 1986, p. 1
[14]Carey, John, Mary Hager, David Lewis, and Sam Seibert, "Can Fuel Cells Light the Way?" *Newsweek,* May 21, 1984, p. 78.
[15]Johnson Matthey, *op. cit.*
[16]*Ibid.,* p. 81.
[17]Free, John, "Fuel Cells - After Twenty Years of Disappointments, They're Coming On-line," *Popular Science,* March 1986, p. 81.
[18]Blackburn, John O., *The Renewable Energy Alternative: How the United States Can Prosper Without Nuclear Energy or Coal,* 1987, p. 11.
[19]"Assessment of Research Needs for Advanced Fuel Cells," *Energy,* January-February 1986, p. X.
[20]*Ibid.*
[21]*Ibid.,* p. XI.
[22]*Ibid.*
[123]*Newsweek, op. cit.*
[24]Reed and Lerner, *Science,* December 28, 1973, p. 1301.
[25]Dworetzky, Tom, "New Life for a Cell: The Ceramic Solution," *Discover,* July 1987, p. 16.

[26]Scott, David, "Stirling Powered Sub," *Popular Science,* January 1988, p. 83.

[27]"The Stirling Engine," NASA, 1986, p. 14.

[28]Samuel Glasstone, *Energy Deskbook,* U.S. Department of Energy, Technical Information Center, pp. 381-385.

[29]NASA, *op. cit.,* p. 12.

[30]Early figures taken from NASA report based on 1978 test results.

[31]Richmond, Jeff, "Stirling Cycles Back," *Popular Mechanics,* June 1985, p. 144.

[32]Shaltens, Richard K., "Comparison Stirling Engines for Use with a 25-KW Dish Electric Conversion System," 22nd Intersociety Energy Conversion Engineering Confrence, Philadelphia 10-14 August 1987.

GLOSSARY

Acid deposition - Sulphur dioxide and nitrogen oxides, given off during the combustion of fossil fuels, become acidic in the presence of water vapor in the atmosphere or similar molecules. They form sulfuric acid and nitric acid gas which can be deposited in a dry form or dissolved in earthbound rainwater (acid rain), causing damage to water bodies and their inhabitant fish populations, forests and other plant life, and buildings.

ARCO - Atlantic Richfield Oil Company

Biomass Fuels - Fuels obtained primarily from plants and plant products. Biomass includes trees and other living plants, crop residues, wood and bark residues, and animal manures.

BTU - British Thermal Unit - the amount of heat required to raise the temperature of one pound of water one degree Fahrenheit under standard conditions of pressure and temperature (equal to 252 calories, 778 foot-pounds, 1055 joules, and 0.293 watt-hours). It is the standard unit for measuring quantity of heat energy.

CAFE - Corporate Average Fuel Economy. A fleet-average measure of automobile fuel efficiency (in miles per gallon) that manufacturers must meet for a particular model-year.

Catalytic converter - An air pollution abatement device used to reduce nitrogen oxide emissions from motor vehicles.

CEC - California Energy Commission

CO_2 - Carbon dioxide - A colorless gas produced when carbon is burned in a sufficient supply of oxygen to complete the reaction $C + O_2 = CO_2$. It is a normal constituent of the atmosphere to the extent of about 0.03 percent by volume, but the extensive burning of fossil fuels since the Industrial Revolution has caused a sharp increase in the natural concentration of CO_2 in the atmosphere.

Coal gasification - A general term used to describe the production of a fuel gas from coal. Most processes react coal with steam to produce an intermediate-BTU gas, followed by water-gas shift and methanation to produce a high-BTU natural gas substitute.

Compression ignition engine - An internal combustion engine in which ignition of the liquid fuel injected into the cylinder is performed by the heat of compression of the air charge.

Compression ratio - The ratio of absolute pressure in the cylinder of an engine after and before compression, e.g., 10:1.

Cryogenic techniques - Techniques involving the use of cooling to extremely low temperatures to keep certain fuels, such as liquefied hydrogen, methane, and propane, in a liquid form.

Department of Energy - DOE - In October 1977, the Department of Energy was created to consolidate the multitude of energy oriented government programs and agencies. Three complete Federal agencies (Federal Energy Administration, Energy Research and Development Administration, Federal Power Commission) were combined with parts of five other Departments and Agencies to create DOE.

Diesel engine - A compression ignition engine (see above) that can burn a wide variety of fuels, ranging from natural gas to heavy distillate fuel oils.

Elastomer - A stretchable substance, resembling rubber that will return to approximately its original length.

Energy efficiency - The efficiency with which energy-consuming technologies make use of different energy sources; e.g., automobile fuel-efficiency; furnace combustion efficiency; electric generation efficiency; etc. Also may refer to energy conservation technologies, such as an energy-efficient home.

Electrode - Reactive materials, such as metals and metal oxides, through which an electric current enters or leaves a medium.

Electrolysis - The chemical decomposition of a substance when electricity is passed through it in solution or in the molten state.

Electron - An elementary particle bearing a negative electric charge. Electrons orbit the atomic nucleus in a chemical element; their transfer or rearrangement between atoms underlies all chemical reactions.

Energy - The capacity to do work. It takes such forms as potential, kinetic, heat, chemical, electrical, nuclear, and radiant energy. Potential energy arises by virtue of the position or configuration of matter. Kinetic energy is energy of motion. Heat energy is the kinetic energy of molecules. Chemical energy is heat produced when atoms combine or separate. Electrical energy arises out of the capacity of moving electrons to produce heat, electromagnetic radiation, and magnetic fields. Nuclear energy is heat formed in the nuclear fission (atom-splitting) or fusion processes. Only radiant heat can exist alone; all other forms require the presence of matter. Some forms of energy can be converted into other forms, and all forms are ultimately converted into heat.

Energy density - Energy per unit of weight contained in a fuel. Energy density can be quite different for the same fuel depending on its state (liquid or vapor).

Energy Research and Development Administration (ERDA) - An independent executive agency of the Federal Government with responsibility for management of research and development in all energy matters. Creation of ERDA was concurrent with dissolution of the Atomic Energy Commission (AEC), and nuclear programs were transferred to ERDA. Now absorbed into the Department of Energy.

Enhanced oil recovery - Increased recovery of crude oil (and natural gas in the case of enhanced gas recovery) from a reservoir, which is achieved by the external application of physical or chemical processes that supplement naturally occurring or simple fluid injection processes.

Environmental Protection Agency (EPA) - A Federal agency created in 1970 to permit coordinated governmental action for protection of the environment by the systematic abatement and control of pollution through integration of research, monitoring, standard setting, and enforcement activities.

Ethanol - Ethyl alcohol or grain alcohol, C_2H_5OH. It is the alcohol contained in intoxicating beverages. Ethanol can be produced from biomass by the conversion process called fermentation.

199

External combustion engine - An engine in which the fuel is burned outside the cylinders; e.g., the Stirling engine.

FERC -Federal Energy Regulatory Commission

FIP - Federal Implementation Plan

Fiscal Year (FY) - Government's 12-month financial year, which starts on October 1 and runs through September 30 of the following calendar year; for example, FY-1989 extends from October 1988 through September 1989.

Fission - The splitting of a heavy nucleus into two approximately equal parts (which are nuclei of lighter elements), accompanied by the release of a relatively large amount of energy and generally one or more neutrons. Fission can occur spontaneously, but usually is caused by nuclear absorption of gamma rays, neutrons, or other particles.

FFV - Flexible Fuel Vehicle.

Fossil fuel - Any naturally occurring fuel of an organic nature such as coal, crude oil, and natural gas.

Fuel cell - Hydrogen is chemically reacted with oxygen from the air to produce electricity in what is essentially a reverse-electrolysis process.

Fusion (nuclear) - A nuclear reaction in which two light atomic nuclei unite (or fuse) to form a single nucleus of a heavier atom. This process takes place in the sun and in other active stars. It has been duplicated by physicists in the hydrogen bomb, though it has not yet been controlled for useful purposes, as has the fission reaction. Recent experiments indicate that fusion reactions may be possible at room temperature.

GAO - General Accounting Office

Geothermal energy - The heat energy available in the rocks, hot water, and steam in the earth's subsurface. Usually used to generate electricity.

Greenhouse effect - Human activity is generating increasing amounts of carbon dioxide and several other gases (primarily methane, chlorofluorocarbons, tropospheric ozone, and nitrous oxide). These gases let energy from the sun pass through the atmosphere but absorb longer wavelength radiation reflected from the earth's surface. Trapped rays are converted into heat in a process similar to that which occurs in a greenhouse, warming the atmosphere beyond historic norms.

Hydrocarbon - HC - Any of a large class of organic compounds composed solely of carbon and hydrogen. The compounds having a small number of carbon and hydrogen atoms in their molecule are usually gaseous; those with a larger number of atoms are liquid and the compounds with the largest number of atoms are solid.

Hydroelectric plant - An electric power plant in which the energy of falling water is converted into electricity by turning a turbine generator.

Hydrogen - The lightest element, No. 1 in the atomic series. It has two natural isotopes of atomic weights 1 and 2. The first is ordinary hydrogen; the second is deuterium, or heavy hydrogen. A third isotope, tritium, atomic weight 3, is a radioactive form produced in reactors by bombarding lithium-6 with neutrons. Chemical symbol H.

Hydrogen economy - Use of hydrogen as an energy source to be piped to local storage or usage areas. The hydrogen would be burned to produce heat, used as a transportation fuel, in industrial processes, or in fuel cells to generate electricity.

Internal combustion - Pertaining to any engine in which the heat or pressure necessary to produce power is developed in the engine cylinder by the burning of a mixture of air and fuel, and converted into mechanical work by means of a piston. The automobile engine is a common example.

Interstate gas - Natural gas which enters interstate commerce (crosses state lines) and hence is subject to Federal price controls. Natural gas sold to pipelines under the jurisdiction of the FERC.

Intrastate gas - Natural gas which is both produced and consumed within the same state. Prior to 1978, was not subject to Federal (FERC) price controls.

Kilowatt - A unit of power equal to 1,000 watts or to energy consumption at a rate of 1,000 joules per second. It is usually used for electrical power. An electric motor rated at one horsepower uses electrical energy at a rate of about 3/4 kilowatt.

Kilowatt-hour (Kwh) - The unit of energy equal to that expended in one hour at rate of 1 kilowatt or 3,413 British thermal units.

Latent heat of vaporization - The quantity of heat required to change a unit weight of liquid to vapor with no change in temperature.

Lifecycle cost - The accumulation of all funds spent for the purchase, installation, operation, and maintenance of a system over its useful life. The accumulation generally includes a discounting of future cost to reflect the relative value of money over time.

Light Water Reactor (LWR) - Nuclear reactor in which water is the primary coolant/moderator with low-enriched uranium fuel. There are two commercial light-water reactor types - the Boiling Water Reactor (BWR) and the Pressurized Water Reactor (PWR).

Liquefaction - Conversion of a solid to a liquid by heat, or from a gas into a liquid by cold or pressure.

Liquefied natural gas (LNG) - Natural gas cooled to -160°C so it forms a liquid at approximately atmospheric pressure. As natural gas becomes liquid, it reduces volume by a factor of almost 600, thus allowing both economical storage and economical long distance transportation in high pressure cryogenic containers, usually on LNG tankers.

M85 - A blend of 85% methanol, 15% gasoline.

M100 - Pure, "neat" (100%) methanol.

Megawatt - A unit of power. A megawatt equals 1,000 kilowatts, or one million watts.

Methanol - A light volatile, flammable, poisonous liquid alcohol, (CH_3OH) formed in the destructive distillation of wood or made synthetically and used especially as a fuel, a solvent, an antifreeze, or a denaturant for ethyl alcohol, and in the synthesis of other chemicals. Methanol can be used as a fuel for motor vehicles.

Methane - A colorless odorless flammable gaseous hydrocarbon that is a product of the decomposition of organic matter in marshes or mines or of the carbonization of coal. It is used as a fuel and as a raw material in chemical syntheses. Chemical formula CH_4. CH_4 is the prime constituent of natural gas. Methane can also be made by certain biomass conversion processes.

Metric ton - A unit of mass and weight that equals 1,000 kilograms or 2,204.6 avoirdupois pounds.

MMBTU - Million Btu's (British thermal units)

NAAQS - National Ambient Air Quality Standard

Natural gas - Naturally occurring mixtures of hydrocarbon gases and vapors, found in porous geologic formations beneath the earth's surface, often in association with petroleum. The more important are methane, ethane, propane, butane, pentane, and hexane. The energy content of natural gas is usually taken as 1,032 British thermal units per standard cubic foot.

Nitrogen dioxide - A compound produced by the oxidation of nitric oxide in the atmosphere; a major contributor to photochemical smog; also the most dangerous of mine gases produced by the incomplete detonation of some explosives. Chemical formula NO_2.

Nitrogen oxides - Compounds of nitrogen and oxygen which may be produced by the burning of fossil fuels. They are very harmful to health and a contributor to the formation of smogs. Chemical formula NO_x.

Nuclear energy - The energy liberated by a nuclear reaction (fission or fusion) or by radioactive decay. When released in sufficient and controlled quantity, this heat energy may be converted to electrical energy.

Octane - A rating scale used to grade gasoline as to its antiknock properties; also any of several isomeric liquid paraffin hydrocarbons, C_8H_{18}; specifically, normal octane, a colorless liquid boiling at 124.6°C found in petroleum.

OPEC - Organization of Petroleum Exporting Countries - Founded in 1960 to unify and coordinate petroleum policies of the members. The members and the date of membership are: Abu Dhabi (1967); Algeria (1969); Indonesia(1962); Iran (1960); Iraq(1960); Kuwait(1960); Libya(1962); Nigeria (1971); Qatar(1961); Saudi Arabia (1960); and Venezuela (1960). OPEC headquarters is in Vienna, Austria.

Ozone - A molecule containing three oxygen atoms (O_3). It occurs in minute quantities in the air near the earth's surface (troposphere) and in larger quantities in the stratosphere as a product of the action of ultraviolet light on oxygen (O_2). At the top of the earth's atmosphere, stratospheric ozone acts as a protective layer by absorbing ultraviolet radiation. Tropospheric ozone is a major component of photochemical smog; it has a sharp unpleasant odor and is an eye irritant.

Petroleum - Hydrocarbon material occurring naturally in the earth and composed predominantly of mixtures of chemical compounds of carbon and hydrogen with or without other nonmetallic elements, such as sulfur, oxygen, nitrogen, etc. Petroleum may contain compounds in the gaseous, liquid, and/or solid state, depending on the nature of these compounds and the existent conditions of temperature and pressure.

Photovoltaic process - Process of converting light rays directly into electricity without going through the intermediate steps involving heat generation and conversion. In photovoltaics, solar cells convert sunlight into a stream of electrons easily converted into household current.

Plutonium - A heavy, radioactive, man-made, metallic element with atomic number 94. It occurs in nature in trace amounts only. Its most important isotope is fissionable plutonium-239, produced by neutron irradiation of uranium-238. It is used for reactor fuel and weapons. Chemical symbol Pu.

Price Anderson Act - Provides insurance for damages arising from a nuclear accident. It requires a utility which operates a nuclear power plant to purchase the maximum amount of liability insurance it can obtain. The federal government provides additional insurance to pay the public for damages from a nuclear accident.

Primary energy - Energy in its naturally-occurring form (coal, oil, uranium) before conversion to end-use forms, such as electricity.

Propane - A gaseous member of the paraffin series of hydrocarbons which, when liquefied under pressure, is one of the components of liquefied petroleum (LP) gas. Contains approximately 2,500 British thermal units per cubic foot. Used for domestic heating and for cooking. Chemical formula C_3H_8.

Proved reserves - The estimated quantity of energy reserves (oil, natural gas, coal, etc.) which geological and engineering data demonstrate with reasonable certainty to be recoverable from known deposits under existing economic and operating conditions.

Prudhoe Bay field - The collection of reservoirs in Alaska's North Slope, proved in 1970, and currently in production. This includes reserves of about 9.6 billion barrels of oil and 33 Tcf of natural gas.

SIP - State Implementation Plan

Solar Energy Research Institute (SERI) - The Solar Energy Research, Development and Demonstration Act of 1974 called for the establishment of a Solar Energy Research Institute whose general mission would be to support DOE's solar energy program and foster the widespread use of all aspects of solar technology, including direct solar conversion (photovoltaics), solar heating and cooling, solar thermal power generation, wing, ocean thermal conversion, and biomass conversion.

Spent fuel - Nuclear reactor fuel that has been used (irradiated) to the extent that it can no longer effectively sustain a chain reaction.

Spent fuel reprocessing - Treatment of spent reactor fuel to recover fissile isotopes (U_{235}; Pu_{237}) and separate-out other fission products.

Stirling engine - An external combustion engine where the heat is used to expand a working gas in the cylinder, which pushes down on the piston to produce the power stroke. The Stirling concept was invented in 1816 by a Scottish parson.

Sulfur oxides - Compounds composed of sulfur and oxygen produced by the burning of sulfur and its compounds such as those occurring naturally in deposits of coal, oil, and gas.

Synthesis gas - A mixture of carbon monoxide (CO) and hydrogen(H_2), containing small amounts of nitrogen, some carbon dioxide, and various trace impurities; prepared for petrochemical synthesizing processes.

Tetraethyl lead (TEL) - A volatile lead compound which is added in concentrations of up to 3 milliliters per gallon to motor and aviation gasoline to increase the antiknock properties of the fuel. Formula (Pb $(C_2H_5)4$.

Thermal efficiency - Relating to heat, a percentage indicating the available British thermal unit input of a fuel that is converted to useful purposes; e.g., the ratio of the electric power produced by a power plant to the amount of heat produced by the fuel; a measure of the efficiency with which the plant converts thermal to electrical energy.

Toluene - A colorless liquid which occurs in coal tar and wood tar. Used as a solvent and as an intermediate for its derivatives. Symbol $C_6H_5CH_3$.

Trans Alaska Pipeline System (TAPS) - An 800-mile pipeline, across the state of Alaska, constructed by the Alyeska Pipeline Service Company of Anchorage, Alaska, for the purposes of making the 9.6 billion barrel reserves at Prudhoe Bay, Alaska, available to U.S. industry and consumers. At capacity, more than two million barrels a day can be transported from Prudhoe Bay to the Port of Valdez.

Trillion cubic feet (Tcf) - A unit of measure commonly used for natural gas; equivalent to 39.3 million tons of coal or 184 million barrels of oil.

Uranium - A radioactive element with the atomic number 92 and, as found in natural ores, an average atomic weight of approximately 238. The two principal natural isotopes are uranium-235, which is fissionable, and uranium-238, which is fertile. Natural uranium also includes a minute amount of uranium-234. Uranium is the basic raw material of nuclear energy. Chemical symbol U.

Vapor pressure - The pressure at which a liquid and its vapors are in equilibrium at a definite temperature. If the vapor pressure reaches the prevailing atmospheric pressure, the liquid boils.

Volatile - Capable of being readily evaporated at a relatively low temperature.

Waste heat - Heat which is at temperatures close to the ambient and hence is not valuable for production of power and is discharged to the environment.

Wind energy - A form of solar energy, since winds are caused by variations in the amount of heat the sun sends to different parts of the earth. Electricity is produced when a windmill catches the wind and revolves, rotating a turbine which powers an electric generator. Major drawbacks of wind energy are the high capital costs and limitations on locations due to the large amount of wind needed to generate the 6,000 watts of electricity used in the average home. As with sunlight-produced (solar heat, photovoltaics) energy, some backup resource is needed.

BIBLIOGRAPHY

Agoos, Alice and Kristine Portnoy. "Methanol Has A New Look: Tight Supply." *Chemical Week.* 18 November 1987: 44-48.

Alexander, R. Jack. "Methanol: America's Answer to Motor Fuel Independence." Presentation to Governing Boards of Automotive Warehouse Distributors Association and Motor Equipment Association. Carlsbad, California: June 1982.

Alcohol Week. 1 August 1983: 2.
            7 September 1987: 8.
            13 February 1984: 6.

"Assessment of Research Needs for Advanced Fuel Cells." *Energy.* 11, January/February 1986: 1-229.

Anderson, Earl. "Lead Cut Gives Alcohol Crack At Gasoline Blend Market." *Chemical & Engineering News.* 8 April 1985: 17-18.

--"Methanol: Chasing An Elusive Fuel Market." *Chemical & Engineering News.* 16 July 1984: 9-16.

--"Methanol Touted As Best Alternate Fuel For Gasoline." *Chemical & Engineering News.* 11 June 1984: 14-16.

"ARCO to sell Methanol Fuel in California." *Chemical & Engineering News.* 1 June 1987: 5.

Aucklund Regional Authority. "A Report on New Zealand Bus Trials Using Ignition Improved Methanol." Auckland, 1987.

Barry, John M. "Congress Pushes Homegrown Energy Again." *Business Month.* November 1987: 70.

Begley, Ronald. "Crude Rise Nudges Olefins Feeds: Petrochemicals '87". *Chemical Marketing Reporter.* 6 April 1987: 29-30.

Belanger, John. "Acid Rain Key Topic of Summer." *Palm Beach Post.* 16 March 1986: A22.

Bern, Reginald I. "Hydrogen Routes' Future Keyed to Economics." *Chemical Engineer.* 14 July 1980: 80-84.

Bernton, Hal, William Kovarik, and Scott Sklar. *The Forbidden Fuel: Power Alcohol in the 20th Century.* New York: B. Griffin, 1982.

Berry, Bryan H. "Auto Energy: Meeting the Need for the Future." *Iron Age.* 24 June 1983: 42-47.

Best, Don. "Solar Cells: Still a Tough Sell." *Sierra.* May-June 1988: 28-29.

Blackburn, John O. *The Renewable Energy Alternative: How the United States Can Prosper Without Nuclear Energy or Coal.* Durham: Duke University Press, 1987.

Bollag, Burton. "Cut Your Switch at Red Light to Save a Tree, Motorists Told." *The Miami Herald.* June 1987: 2A.

California Energy Commission. *Methanol One Fact Sheet.*

Carey, John, Mary Hager, David Lewis, and Sam Seibert, "Can Fuel Cells Light the Way?" *Newsweek.* 21 May 1984: 78.

"Clean Air Advocates: Still (Wheezing, Gasping, Crying) Trying After All These Years." *Sierra.* September-October 1987: 13.

"Clean Air Act Mandate For Alcohol Fuels Passes Senate." *Alcohol Week.* 3 August 1987: 1, 5.

Cohen, Laurence H. and Herman L. Muller. "Methanol Cannot Economically Dislodge Gasoline." *Oil & Gas Journal.* 28 January 1985: 119-123.

Congress of the United States. *Curtailing Acid Rain: Cost, Budget, and Coal-Market Effects.* Washington: Congressional Budget Office, June 1986.

"Continuous Fuel Cell Research Hastens Future Commercialization of A Non-Polluting Energy Source." *Johnson Matthey Newsline.* January 1986: 1-2.

D'Alessandro, Bill. "Dark Days for Solar." *Sierra.* July-August 1987: 37.

"Dawn Bankruptcy May Spell End to Farmer's Home Administration Ethanol Loan Guarantee Program." *Alcohol Week.* 23 November 1987: 3.

Deudney, Daniel and Christopher Flavin. *Renewable Energy: The Power to Choose.* New York: W. W. Norton, 1983.

Dixon, R.H. "Methanol: U.S. and Global Picture." Presented at Oxygenated Fuel Conference, Arlington,VA. 18 November 1982.

Doig, Stephen K. "Future Shock: Burger Boxes, Cars, Even the Fridge are Warping Our Climate." *Miami Herald.* 11 October 1987: 1, 16.

Dworetzky, Tom. "New Life For A Cell: The Ceramic Solution." *Discover.* July 1987: 16.

"EC Bid to Delay BioEthanol Subsidy Sends Industry Reeling in Europe." *Alcohol Week.* 23 November 1987: 7.

Economist. 8 August 1987: 89.

19 October 1985: 103-104.

Emond, Mark. "Is Methanol Niche Shaping Up In The West?" *National Petroleum News.* August 1987: 18-19.

"EPA Clears Way for More Alcohol Fuels". *Chemical & Engineering News.* 3 November 1986: 23.

"EPA Says Fuel Methanol Is Only Practical Way to Reduce Ozone Pollution." *Alcohol Week.* 16 November 1987: 1, 3.

"EPA Volatility Regulations May Favor MTBE Over Ethanol to Replace Butanes." *Alcohol Week.* 18 August 1987: 1.

"Estimate Cut for U.S. Potential Gas Resources." *Oil & Gas Journal.* 27 April 1987: 105.

Federal Highway Administration. "Section 152 Study: Methane Conversion for Highway Fuel Use." July 1986 (Draft).

Flavin, Christopher. "Reassessing Nuclear Power: The Fallout From Chernobyl." Washington: Worldwatch, 1987.

Flavin, Christopher. "Nuclear Power: The Market Test." Washington: Worldwatch, 1983.

Flavin, Christopher, and Alan B. Durning. "Building On the Age of Energy Efficiency." Washington: Worldwatch, 1988.

"Flexible Fuel Vehicle Could Help Benefits of Methanol." *Research & Development.* August 1986: 33.

Fox, Michael R. "Future Energy Supplies." *Vital Speeches.* 30 March 1985: 554-5.

Fraas, Arthur, and Albert McGartland. "Alternate Fuels for Pollution Control: An Empirical Evaluation of Benefits and Costs." Western Economic International Association Meetings, Alternate Fuels Session. 30 June-3 July 1988.

Free, John. "Fuel Cells: After 20 Years of Disappointments, They're Coming on Line." *Popular Science.* March 1986: 80-82.

"Gasohol Finds Renewed Life As Substitute for Leaded Fuel." *PB Post Evening Times.* 17 October 1985: A18.

Glasstone, Samuel. *Energy Deskbook.* U.S. Department of Energy, Technical Information Center. 1982.

Gottschalk, Earl. Jr. "WPPSS Underwriters Pact Could Spur Settlement With Other Major Defenders." *Wall Street Journal.* 14 November 1987: 12.

Gray, C. Boyden. "Octane, Ozone, and Obstinacy," *Regulation.* 1987.

Gray, Charles L. and Jeffrey A. Alson. *Moving America to Methanol.* Ann Arbor: University of Michigan Press, 1985.

"Greenhouse Effect is Real, Scientists Say." *Palm Beach Post.* 11 June 1986: 5A.

"Gulf Tankers May Have to Pay Escort Fees." *Miami Herald.* 15 October 1987: 4.

Haggin, Joseph. "World Methanol Situation Poses Challenge in Process Design." *Chemical & Engineering News.* 16 July 1984: 31-35.

--"Methanol From Biomass Draws Closer to Market." *Chemical & Engineering News.* 12 July 1982: 24-25.

"High MTBE Sales Lower Colorado's Carbon Monoxide Reduction Figures." *Alcohol Week.* 1 February 1988: 1,9.

Hiro, Dilup. "Moscow's Double Dealing in the Gulf." *Wall Street Journal.* 30 July 1987: 22.

Hirsch, Robert L. "Impending United States Energy Crisis." *Science.* 20 March 1987: 1470-71.

Hoffmann, Peter. *The Forever Fuel: The Story of Hydrogen.* Boulder: Westview, 1987.

--"Fueling the Future With Hydrogen." *The Washington Post.* 6 November 1987: B-3.

"House Approves Sharp Methanol Bill By Margin Despite Veto Threat." *Alcohol Week.* 21 December 1987: 1-2.

Hunt, V. Daniel. *Energy Dictionary.* New York: Van Nostrand Reinhold Co., 1979.

Information Resources, Inc. "Understanding the Challenges and Future of Fuel Alcohol in the United States." Prepared for U.S. Department of Energy, Office of Alcohol Fuel, Washington, D.C., September 1988.

Ingrassia, Lawrence. "PS of NH Bankruptcy Plea Marks End to Era, But May Affect Industry Little." *Wall Street Journal.* 1 February 1988; 60.

Jack Faucett Associates. "Methanol Prices During Transition." 12 November 1986.

Jackson, M.D. Powers, C.A. and Fong, D.W. "Methanol Fueled Transit Bus Demonstration." Paper no 83-DGP-2. American Society of Mechanical Engineers.

Jonchere, R. "Methanol Seen As Hydrogen Source." *Oil & Gas Journal.* 14 June 1976: 71-73.

Keefe, Lisa M. "Natural Gas." *Forbes.* 13 January 1986: 180-182.

Kahn, Helen. "Methanol Trendy Among Lawmakers." *Automotive News.* 2 December 1985: 8.

--"Methanol Becoming A Pet in Congress, But Problems Loom." *Automotive News.* 9 April 1984: 2.

Kissinger, Henry A. Foreword. *The Critical Link: Energy and National Security in the1980's.* A Report of the Energy, Natural Resources and Security Studies Division. Cambridge: Ballinger, 1982: xv-xxii.

Klein, Dale. "For the Lack of An Energy Policy." editorial. The *Houston Chronicle.* 20 August 1987: 27.

"L.A. to Fight Dirty Air With Dramatic Curbs on How People Live." *Atlanta Constitution.* 18 March 1989, 1A.

"Lack of Methanol Vehicle Standards Slows Production For Consumers." *Alcohol Week.* 30 November 1987: 4-5.

Larsen, Robert P. and D.J. Santini. "Rationale for Converting the U.S. Transportation System to Methanol Fuel". Chicago: Argonne National Laboratory, 1986.

Lemonick, Michael. "Kicking the Gasoline Habit." *Science Digest.* May 1984: 29.

Lincoln, John Ware. *Methanol and Other Ways Around the Gas Pump.* Charlotte, Vt: Garden Way, 1976.

Long, Frank W. "Methanol". *U.S. Petrochemicals: Technologies, Markets and Economics.* ed Arthur M. Brownstein. Tulsa: Petroleum Publishers, 1972: 81-92.

Marsden, S.S. "Methanol As A Viable Energy Source in Today's World." *Annual Review of Energy.* 8 (1983): 333-354.

McGill, Ralph N. and E. Eugene Ecklund. "Introducing Methanol-Fueled Vehicles Into Government Fleet Operations." Proceedings of the VII International Symposium on Alcohol Fuels. 20-23 October 1986. Paris.

Melloan, George. "Californians Will Pay Dearly For PURPA Power." *Wall Street Journal.* 31 March 1987: 37.

"Methanol Fueled Bus Brings Clean Transit." *Design News.* 19 December 1983: 15.

"Methanol Prices Soar in U.S. Gulf As Supplies Tighten Around the World." *Alcohol Week.* 4 January 1988: 1-2.

"Methanol Supply Seen Shy of Demand in Early 1990's." *Oil & Gas Journal.* 18 August 1986: 49.

Paul, J.K. ed. *Methanol Technology and Application in Motor Fuels.* Park Ridge, N.J.: Noyes Data Corp, 1978.

Mills, G. Alex and E. Eugene Ecklund. "Alcohols As Components of Transportation Fuels." *Annual Review of Energy.* 12 (1987): 47-80.

National Acid Precipitation Assessment Program. *NAPAP Interim Assessment* (4 volumes), 1987.

Nesmith, Jeff. "Increasing CO_2 Levels Worry Scientists." *Palm Beach Post.* 7 September 1986.

"The New Route to CO from Methanol - Via Water." *Chemical Week.* 10 October 1984: 48-55.

"The Next Oil Crisis." *Commonweal.* 3 March 1987: 132.

"Nuclear Liability Measure is Approved by the Senate." *Wall Street Journal.* 8 August 1988: 5.

Nulty, Peter. "Guess Which Fuel Is Looking Hot." *Fortune.* 6 June 1987: 94-102.

O'Hare, T.E., R.S. Sapienza, D. Mahajan, and G.T. Skaperdas. "Methanol for Transportation of Natural Gas Values." Presented at American Society of Mechanical Engineers, Joint Methanol Conference. Columbus, OH, 25-27 June 1986.

Olsen, Walter. "Dirty Coal, No Clear Air." *Barrons*. Editorial. 30 November 1987: 11.

Othmer, Donald F. "Methanol Fuel for Automobiles." *Chemical Engineering Progress*. October 1985: 116-21.

Owen, David. "Octane and Knock." *The Atlantic Monthly*. August 1987.

"The Ozone Hole." *Scientific American*. July 1987: 20.

Pasternak, Alan. "Methyl Alcohol - A Potential Fuel for Transportation." *The Energy Technology Handbook*. New York: McGraw Hill, 1977: 4-45 to 4-49.

Pasztor, Andy. "U.S. Wants to Combat Gasoline Fumes from Cars Refueling at Service Stations." *Wall Street Journal*. 17 July 1984.

Paul, Bill. "Ex-NRC Gadfly Returns to Sting Utilities." *Wall Street Journal*. August 1987.

Pollock, Cynthia. "Decommissioning: Nuclear Power's Missing Link." Washington: Worldwatch, 1986.

"Pollution Strengthens Hurricanes." *The Miami Herald*. 2 April 1987: 13A.

Pope, Christopher. "Methanol Future Planned." *Renewable Energy Future*. June 1984: 1, 6.

Postel, Sandra. "Altering the Earth's Chemistry: Assessing the Risks." Washington: Worldwatch, 1986.

Rasheed, Vic. "Gasohol Is A Waste of Energy. *Wall Street Journal*. editorial. 29 October 1987.

Reed, T.B. and R.M. Lerner. "Methanol: A Versatile Fuel for Immediate Use." *Science*. 28 December 1973: 1299-1304.

Richmond, Jeff. "Stirling Cycles Back." *Popular Mechanics*. June 1985: 143-150.

"Rise in Sea Levels Threatens Coasts." *Palm Beach Post Evening Times*. 27 May 1986: 4A.

"Rockefeller Calls CAFE Credit Caps 'Key' To Passing Methanol Bill In Senate." *Alcohol Week.* 14 December 1987: 11.

Roeder, Bill. "Piping Methanol From Alaska." *Newsweek.* 18 May 1987: 37.

Rogers, David. "Major Changes in Nuclear Waste Plan Approved by Senate Panel for Fiscal 1988." *Wall Street Journal.* 16 September 1987: 12.

Santini, D.J. "The Petroleum Problem: Managing the Gap." Illinois: Argonne National Laboratory, 1986.

Scheibla, Shirley Hobbs. "Unwise At Any Speed: Those Absurd Gasoline Mileage Standards Should Be Scrapped." *Barrons.* editorial. 4 August 1986: 9.

"Scientist Sees Drastic Greenhouse Effect by 2020." *Palm Beach Post.* 11 June 1986: 5A.

Scott, David. "Stirling-Powered Sub." *Popular Science.* January 1988: 82-83.

Segal, Migdon R. "Alcohol Fuels" CRS Issue Brief No. IB 74087, Congressional Research Service, 26 May 1989.

Segal, Migdon R., A. Barry Carr, *et al.* "Analysis of Possible Effects of H.R. 2052, Legislation Mandating Use of Ethanol in Gasoline." CRS Report for Congress, No. 87-819 SPR, 13 October 1987.

Segal, Migdon R., A Barry Carr, *et al.* "Analysis of Possible Effects of H.R. 2031, Legislation Mandating Use of Ethanol and Methanol in Gasoline." CRS Report for Congress, No. 88-71 SPR, 17 November 1987.

Shaner, J. Richard. "Methanol Motor Fuel Looking Good Again." *National Petroleum News.* July 1987: 14.

"Sharp Methanol Bill Amendment Seeks Government FFV Sales To Public." *Alcohol Week.* 23 November 1987: 5-6.

Shaw, Robert W. "Air Pollution By Particles." *Scientific American.* July 1987: 96-100.

Shea, Cynthia Pollock. "Renewable Energy: Today's Contribution, Tomorrow's Promise." Washington: Worldwatch, 1988.

Simonietti, Tony. "Methanol: Classic Fuel Fits the Future." *GM Today.* 11 October 1985: 2, 7.

Smith, Kenneth D., Dan W. Wong, Don S. Kondoleon, Cindy A. Sullivan. "The California State Methanol Program-Creating a Market." and "Methanol As Ozone Control Strategy in the Los Angeles Area." Presented at the VI International Symposium on Alcohol Fuels Technology, Ottawa, Canada. 21-25 May 1984.

Solis, Diana. "Texas Nuclear Plant Enters Critical Stage." *Wall Street Journal.* 8 March 1988: 6.

Solomon, Caleb. "Getting Hooked: As U.S. Wells Go Dry, Reliance on Oil Imports is Sure to Keep Rising." *Wall Street Journal.* 30 March 1988: 1,8.

Stamner, Larry B. "AQMD Oks Plan to Cut Use of Diesel Gasoline." *Los Angeles Times.* 9 January 1988: I-1.

Starke, Linda, ed. *State of the World 1988.* A Worldwatch Report on Progress Toward A Sustainable Society. New York: W.W. Norton, 1988.

Stinson, Stephen C. "Methanol Primed for Future Energy Role." *Chemical and Engineering News.* 2 April 1979: 28-30.

"The Stirling Engine: Mod II Design Report." NASA, Lewis Research Center, 1986.

Stobaugh, Robert and Daniel Yergin, eds. *Energy Future: A Report of the Harvard Business School, Energy Project.* New York: Random House, 1979.

Strand, Robert. "High Oil Prices, Tax Breaks Fueled Windmill Industry." Miami Herald. 25 January 1987: 3C.

"Study: Pollution Warming Climate, Destroying Farms." *Palm Beach Post.* 20 July 1986: 8A.

Sundstrom, Geoff. "Ford, GM Report On Methanol Cars." *Automotive News.* 29 July 1986: 31.

218

Tanner, James. "New Harvard Study Promises to Revive U.S. Debate with Call For Oil-Import Fee." *Wall Street Journal.* 21 September 1987: 4.

--"Panic After Any Gulf Oil Cutoff Wouldn't Last Long, Analysts Say." *The Wall Street Journal.* 23 July 1987: 26.

--"Cartel's Comeback? By Early 90's OPEC Dominate Oil Market." *Wall Street Journal.* 21 November 1987: 15.

Taubes, Gary, with Allen Cheu. "Made in the Shade? No Way." *Discover.* August 1987: 62-71.

Taylor, Robert E. "Oil, Auto Industries Square Off in Attempt to Deflect Burden of Meeting Gas Fumes Plan." *Wall Street Journal.* 30 June 1986: 56.

--"Methanol's Advocates Spark Interest in the Fuel But Critics Say Potential Benefits Will Vaporize." *Wall Street Journal.* 4 September 1987: 38.

--"U.S. Plans to Buy 5,000 Vehicles That Use Methanol." *Wall Street Journal.* 15 July 1987: 2.

--"EPA Deals Blow to Auto, Truck Makers in Ruling on Handling Gasoline Fumes." *Wall Street Journal.* 23 July 1987: 16.

--"Policy Makers, Spurred by Ozone Treaty, Consider Tackling Greenhouse Effect." *Wall Street Journal.* 17 September 1987:38.

The White House, Office of the Press Secretary. "Fact Sheet: President Bush's Clean Air Plan." 12 June 1989.

--"Remarks by the President at the University of Nebraska. 13 June 1989

Toepe R.R., J.E. Bennethum, R.E. Heruth. "Development of Detroit Diesel Allison 6V-92TA Methanol Fueled Coach Engine." SAE Paper No. 831744.

United States Department of Energy. "Assessment of Costs and Benefits of Alternate Fuels in the U.S. Transportation Sector." Washington: GPO, January 1988.

United States Department of Energy, Energy Information Administration. *Monthly Energy Review*, October 1988.

United States Department of Energy. Energy Research Advisory Board. Notes taken from the quarterly meeting. November 1987.

United States Department of Energy. *Energy Security: A Report to the President of the United States.* Washington: GPO, 1987.

United States Department of Energy. *United States Energy Policy, 1980-1988.* Washington: U.S. D.O.E., 15 November 1988.

United States Environmental Protection Agency. "Air Quality Benefits of Alternative Fuels." Washington: GPO, June 1987.

--"Cost and Cost Effectiveness of Alternate Fuels." Washington: GPO, 14 July 1987.

United States General Accounting Office. "Alternative Fuels: Information on DOE's Methanol Vehicle Demonstration Program." Washington: GPO, 1987.

United States House of Representatives, Committee on Energy and Commerce, Subcommittee on Fossil and Synthetic Fuels. Hearings on "Methanol--Fuel of the Future". 99th Cong., 1st session. Washington: GPO, 1986.

United States House of Representatives, Committee on Energy and Commerce. Subcommittees on Fossil and Synthetic Fuels, and Energy Conservation and Power. Hearings on "Methanol As Transportation Fuel". 98th Cong., 2nd session. Washington: GPO, 1984.

United States Office of Technology Assessment. "Acid Rain and Transported Air Pollutants: Implications for Public Policy" Washington: GPO, 1984.

"U.S. Fuel Alcohol Demand to Increase Thru 1990." *Alcohol Week.* 9 November 1987:9.

Vaughn, Eric. "Europe's Wine Surplus and the Ethanol Trade to the Western Hemisphere." 1988 Alcohol Week Conference on Oxygenated Fuels. Frankfurt. 22 June 1988.

Wan, Edward I., Joseph D. Price, John A. Simons. "Methanol Production from Indigenous Resources in New York State: Executive Summary." Albany: NYSERDA, May 1983.

Wartzman, Rick. "Coal Industry Lobby Pits 'Clean Coal' Project Against Pending Rewrite of Acid Rain Legislation." *Wall Street Journal.* 17 September 1987: 70.

Wells, Ken. "Clouded Outlook: As A National Goal, Renewable Energy Has An Uncertain Future," *Wall Street Journal.* 13 February 1986: 1, 18.

Wentworth, Theodore O. and Donald F. Othmer. "Producing Methanol For Fuels." *Chemical Engineering Progress.* August 1982: 29-35.

Wessell, David. "Study in Contrasts: Pilgrim and Millstone, Two Nuclear Plants Have Disparate Fates." *Wall Street Journal.* 28 July 1987: 1, 14.

Williams, Bob and Michael Obel. "Air Quality Concerns Buoy Hopes for U.S. Makers of Alcohol Fuels." *Oil & Gas Journal.* 9 February 1987: 13-16.

Wilson, Carroll, L., Project Director. *Coal - Bridge to the Future: A Report of the World Coal Study.* Cambridge: Ballinger, 1980.

Wilson, Carroll, L. Project Director. *Energy: Global Prospects 1985-2000. A Report of the Workshop on Alternative Energy Strategies.* New York: McGraw Hill, 1977. ix-xi.

Wirth, David. "Dumping in the Atmosphere Changing Climate on Earth." *Palm Beach Post.* 7 September 1986: 9E.

World Commission on Economic Development. Energy 2000: A Global Strategy for Sustainable Development. New Jersey: WCoED, 1987.

ABOUT THE AUTHORS

John H. Perry, Jr.

John H. Perry, Jr., is Chairman of two Florida corporations headquartered in Riviera Beach, Florida: Perry Oceanographics, Inc. and Perry Offshore, Inc. The latter is the largest commercial builder of robot and manned submarines in the U.S. Mr. Perry is Chairman of the Perry Foundation, Inc., a not-for-profit corporation.

Mr. Perry is the former President and Chairman of Perry Publications, Inc. which until 1969, operated 28 newspapers in Florida; *All-Florida* Magazine, a Sunday supplement; *Palm Beach Life* Magazine; the statewide All-Florida News Service, and numerous commercial printing plants in Florida. Perry Publications pioneered the development of computerized printing of newspapers in the early 1960s. He is a Director of the Nassau Guardian (1844) Limited, Nassau, Bahamas.

From 1966 to 1968, he was a member of President Johnson's U.S. Commission on Marine Sciences, Engineering and Resources which, after two years of study, submitted its report entitled, "Our Nation and the Sea." He served as Chairman of the Commission's technology panel. As President of the Bahamas Undersea Research Foundation, Mr. Perry facilitates marine research at the Caribbean Marine Research Center. Through the Perry Energy Systems Division of Perry Oceanographics, he has developed an energy process for making methanol from sea water.

Mr. Perry was born in Seattle, Washington, January 2, 1917; was graduated from Hotchkiss in 1935, Yale in 1939 and attended the Harvard School of Business Administration.

Christiana P. Perry

Christiana Perry graduated *magna cum laude* from Duke University in 1987. She currently is in her second year at Georgetown University Law Center, where she is studying environmental law.

223